The Organic BACKYARD VINEYARD

The Organic BACKYARD VINEYARD

A Step-by-Step Guide to Growing Your Own Grapes

Tom Powers

Timber Press
Portland London

Photo and illustration credits appear on page 186.

This revised, expanded work incorporates portions of *Vineyard Simple: Howto Build and Maintain Your Own Vineyard* by Tom Powers (Alhambra Valley Publications, 2002).

Published in 2012 by Timber Press, Inc.
The Haseltine Building
133 S.W. Second Avenue, Suite 450
Portland, Oregon 97204-3527
timberpress.com

2 The Quadrant
135 Salusbury Road
London NW6 6RJ
timberpress.co.uk

Design by David Jacobson

Printed in China

Library of Congress Cataloging-in-Publication Data
Powers, Tom 1940–
The organic backyard vineyard: a step-by-step guide to growing your own grapes / Tom Powers.—Rev. and expanded.
p. cm.
Includes biographical references and index.
ISBN 978-1-60469-285-3
1. Grapes. 2. Organic viticulture. I. Title. II. Title: Step-by-step guide to growing your own grapes.
SB388.P677 2012
634.8—dc23

2011041688

A catalog record for this book is also available from the British Library.

FRONTISPIECE: Planting and caring for your own vineyard offers a sweet reward: freshly harvested, premium-quality grapes ready to be made into wine.

Dedicated to my family
past and present

Contents

PREFACE 8

ACKNOWLEDGMENTS 10

1 **THE ORGANIC VINEYARD** 13

Why Do You Need a Vineyard? →→ How Grapes Grow →→ The Organic Vineyard →→ Your Location →→ Assessing the Soil

2 **PLANNING YOUR VINEYARD** 37

Choosing your Vines and Rootstocks →→ Popular Grape Varieties →→ Choosing Rootstocks →→ The Trellis System →→ Vineyard Irrigation

3 **INSTALLING YOUR VINEYARD** 63

Installation Timeline →→ Clearing the Land →→ Preparing the Soil →→ Laying out the Vineyard →→ Tools and Equipment →→ Building the Trellis →→ Post-Planting Care

THE VINEYARD YEAR 84

4 **PRUNING AND TRAINING** 91

All About Pruning →→ First Year →→ Second Year →→ Third Year →→ Fourth Year →→ Fifth Year and Beyond →→ Managing Excess Vigor

5 **MAINTAINING THE VINEYARD** 107

Maintenance through the Year →→ Irrigation →→ Fertilizing →→ Organic Fertilizers →→ Weeding and Cover Crops

6 **DISEASES AND PESTS** 125

Integrated Vineyard Management →→ Safe Use and Disposal of Pesticides →→ Organic Disease Control →→ Grapevine Diseases →→ Vineyard Pests →→ Insects in the Vineyard

7 **HARVEST TIME** 159

When Should I Harvest? →→ Balancing Sugar and Acid →→ Harvesting Your Grapes →→ Storage and Winemaking Prep

GLOSSARY 170

HELPFUL CONVERSIONS 176

RESOURCES 177

INDEX 178

PHOTO AND ILLUSTRATION CREDITS 186

LEFT: My father's family in a Florida orange grove around 1911. **RIGHT:** My grandfather Thomas W. Mabrey, on his California dairy farm in the early 1920s, on his way to the Golden State Milk Products Company.

Preface

Vineyards have been around for a very long time. Maps of Mesopotamia, ancient Egypt, and Gaul all show vineyards. The Babylonian legend *The Epic of Gilgamesh* refers to wine as far back as 6000 years ago. The Romans were so fond of wine that they had a god, Bacchus, devoted solely to the fermented juice of the grapevine. In fact, the Romans did a great deal to advance winemaking in Europe, a tradition that was taken up by the French, Italians, and Germans. Early Americans were quick to plant vineyards in the new world, notably Thomas Jefferson in Virginia. Nowadays, there are thousands of vineyards in almost every state and province of North America.

Since the 1970s, winemaking in both Europe and America (and many other regions around the world) has become a science as well as an art, with plenty of opinions and theories about all aspects of growing grapes and making wine. It's true that for premium quality wine, you need good grapes. But many growers and winemakers complicate the process of grape-growing.

As a partner in a vineyard planting and maintenance operation, I have directed the development of more than 100 vineyards, many of them small backyard operations. This first-hand experience has taught me that it is possible to grow grapes organically, provided you work with the natural resources at hand rather than trying to fight against them. My purpose in writing this book is to simplify the process so that an individual with the right conditions can build and manage a vineyard in a sustainable fashion to produce premium quality grapes.

I come from a long line of farmers. My maternal grandfather, Thomas W. Mabrey, settled in the Imperial Valley of California on the Mexican border, raising cotton in 1911 and later starting a dairy farm that supplied milk to the Golden State Creamery. My paternal grandparents, Thomas W. Powers and

Ida Powers, were orange growers in Florida starting in 1910. I can think of no better life than that of a grower caring for the land and its bounty.

The chapters in this book follow roughly the same step-by-step sequence that you will follow in building your vineyard, from the planning stages, through the installation process, and finally to harvest time.

Throughout the book I refer to one scientific process—photosynthesis—which is the means by which the green tissues of plants use energy from the sun to convert carbon dioxide (CO_2), an atmospheric gas, to sugar. This is important because the sugars are the basic building blocks of the components that ultimately give wine its flavor.

Much of the work in the vineyard is devoted to maximizing photosynthesis. That's why proper management of the green canopy of the plants is the most important issue that I can stress in caring for your vines.

Yet please remember that grapevines are very forgiving. If you have the right amount of sun, water, soil, and other basic requirements, even if you make a mistake and pinch or break off the wrong shoot, leaf, or berry, don't worry. There will almost always be one that grows back. Be patient with your vines, take care of your soil, and practice good vineyard management, and your vineyard will reward you with good-quality fruit year after year.

Planting a vineyard is a dream for many people who have even a small amount of land. I hope this book helps you to grow your own vineyard, and that the experience is enormously satisfying, only rarely frustrating, and never boring.

Enjoy.

Acknowledgments

I would like to express my gratitude to the staff at Alhambra Valley Ranch. Martin Martinez, Raul Luna, Damin Luna, Daniel Luna, and Julio Jarquin have been valuable resources in advancing organic practices to allow me to produce grapes that make truly premium wine. Their attention to detail, and the practical solutions they have developed to deal with daily tasks has allowed me the flexibility to explore creative solutions to sustainable farming.

I especially want to acknowledge my ancestors, including my mother, Ruth Powers, my grandfather, Thomas W. Mabrey, and my great grandfather, Anthony Powers. They gained insight into how to care for plants by toiling in small farms in difficult conditions. Their perseverance was an inspiration to me.

Finally I want to acknowledge and dedicate this book to my immediate family: Donna Powers, my wife, and my children Patrick, Jason, and Alicia. You have encouraged me to keep farming in an urban environment. Thank you for all your support.

Tom Powers
Alhambra Valley

WHY DO YOU NEED A VINEYARD?

HOW GRAPES GROW

THE Organic VINEYARD

YOUR LOCATION

ASSESSING THE SOIL

Vines planted on a trellis system are more productive and easier to care for than those simply planted in the ground with no support.

WHY DO YOU NEED A VINEYARD?

If you are simply hoping to plant some table grapes to enjoy for home consumption, you do not need a vineyard. You can grow grapevines up an arbor, over a fence, or against a wall. However, if you want to make wine from your grapes—or to sell them to a winemaker—you will need to train your grapevines on a trellis system. Why is this?

Grapevines planted in rows on a trellis work to your advantage in several ways. First, you have a structure on which you can train your vines to grow in a particular way. This structure allows you to observe the growth of your vines and readily spot any problems or pests. The trellis also allows for easy maintenance, and you can customize it to the height of those who will be doing the most work in the vineyard. Finally, straight rows are the ideal layout for using any kind of mechanical equipment in the vineyard, such as a mower or tractor.

It's not just for your benefit that you need a properly planned and trellised vineyard. Your grapes also benefit. A trellis with an irrigation system makes it possible to provide even and consistent watering and fertilization, which promotes healthy growth. And perhaps most importantly, the vertical trellis is the vehicle that allows the sun to reach the leaves, which is essential for producing good fruit.

Making Wine

First you probably want to know how much wine you can expect to make from your vineyard. Although it's true that even a small suburban lot can house a vineyard, there is not much point in going to the trouble and expense of constructing one if it will not produce enough fruit to meet your needs. The table provides a basic idea of how much land you will need to produce your own wine. Space requirements will vary depending on your vine spacing and the variety, and the productivity of your vines, but it will give you a good general guide.

Keep in mind that your yields (the amount of grapes you can expect to harvest, measured in tons per acre) will also depend on your climate and how well you manage the vineyard. Also note that organic vineyards typically have lower yields than non-organic vineyards, and that yields can vary from year to year.

If you want to have a home vineyard, lack of space doesn't have to stop you. Even a few rows of vines can produce enough grapes to make several hundred bottles of wine every year.

VINEYARD YIELDS

SIZE OF VINEYARD	YIELD (TONS PER ACRE)	GALLONS OF WINE	BOTTLES OF WINE
1 acre	3 (low)	360	1800
1 acre	6 (medium)	720	3600
1 acre	9 (high)	1080	5400
1/2 acre	3 (low)	180	900
1/2 acre	6 (medium)	360	1800
1/2 acre	9 (high)	540	2700
1/4 acre	3 (low)	90	450
1/4 acre	6 (medium)	180	900
1/4 acre	9 (high)	270	1350
1/16 acre	3 (low)	45	225
1/16 acre	6 (medium)	90	450
1/16 acre	9 (high)	135	675
1 row (100 ft. at 5-foot vine spacing)	3 to 9	20 to 35	100 to 175
2 rows (100 ft. at 5-foot vine spacing)	3 to 9	40 to 70	200 to 350

HOW GRAPES **GROW**

The most crucial need that grapevines have is sunlight. Like most plants, grapes need this sunlight in order to undergo photosynthesis. This is the process by which the green tissues of the plant (primarily the leaves) absorb energy from the sun to convert carbon dioxide (CO_2) to sugar (which could be called sugar production). The vine then stores this sugar in its woody tissue in the form of carbohydrates, or starch. The following year, these carbohydrates fuel the growth of the vine and the development of fruit.

Not only is sugar one of the flavor factors in fruit, but it is also sugar that converts to alcohol during fermentation of the fruit. Unless you have a vineyard that allows for a good photosynthetic process to take place, your grapes will not have sufficient sugar levels to make good wine.

Once you understand how to plant and train your vines in such a way that the leaves get maximum sunlight, you can enhance the photosynthetic capacity of your plants. This means, first of all, that nothing can be shading the vineyard, including an adjacent row of vines. It also means that you do not want too many leaves or excessive vigor so that the outer leaves shade the inner leaves. Shaded inner leaves do not do

Photosynthesis and Transpiration

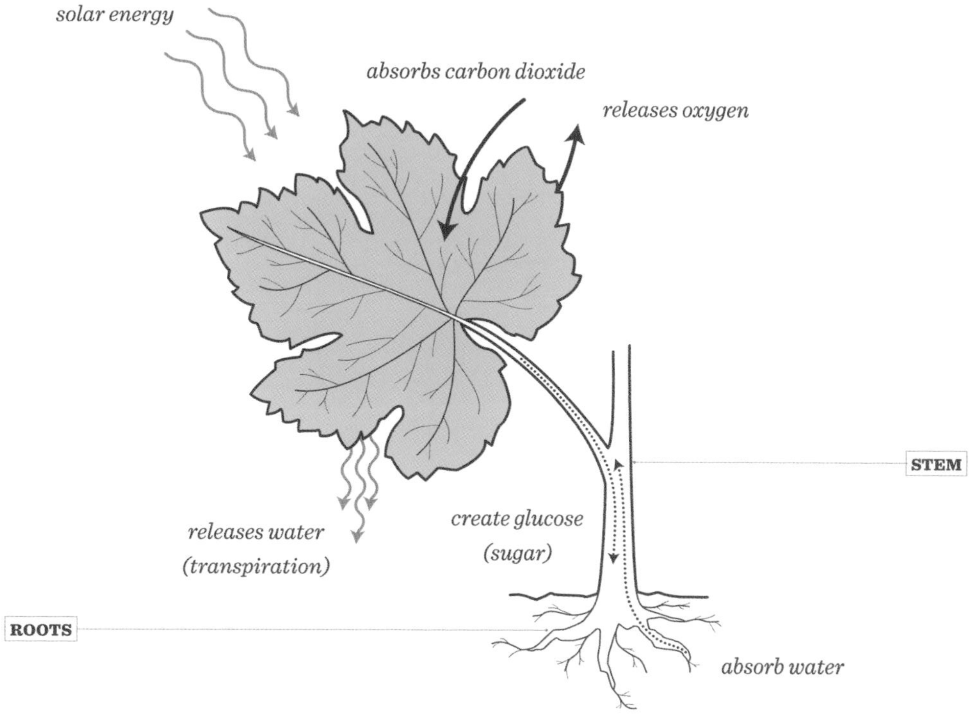

any good and, in fact, are a drain on the plant's stored energy. What you will ultimately want to develop is an open canopy, which will allow the most sun to reach the most leaves. This open canopy has other benefits, including improved air circulation that helps to prevent fungal diseases, which are prevalent in most vineyards worldwide.

Grapevines also have specific temperature requirements. Grapes do not grow in the desert because temperatures above 95°F (35°C) shut down the photosynthetic process. If temperatures are less than 50°F (10°C), photosynthesis will also be inhibited. Somewhere between these two extremes is the place to be. However, warm winters are not necessary to have a vineyard. Grapevines are deciduous plants that go dormant in the winter. Cold weather, ice, and snow do not have a negative effect on vines, except in cases of extreme winter lows.

Grapevines also need water. As the roots absorb water, the moisture is drawn into the leaves where it evaporates in a process known as transpiration. On the underside of each leaf, there are microscopic pores called stomata, which evaporate the water drawn up from the root system. Daytime sunlight induces the pores to open; at night they are closed. The rate of transpiration is closely linked to the process of sugar production.

Interior leaves in dense canopies may be exposed to such low light levels that their stomata do not completely open. So it stands to reason that leaves that are shaded have lower rates of transpiration, less photosynthesis, and thus less sugar production. Exterior leaves on a grapevine canopy are exposed to higher sunlight levels and temperature, and thus transpire more than shaded interior leaves.

Maintenance of the vineyard is primary in order to encourage photosynthesis. I will give you practical steps to take to allow the photosynthetic process to continue at the highest possible rate. Remember, maximum photosynthesis creates maximum sugar, the best fruit, and thus the best wine.

Carefully managing the canopy—the leaves that grow above the fruiting zone—maximizes the amount of sunlight that reaches the leaf surfaces and ensures the maximum possible amount of photosynthesis. It also allows for better management of mildew and other fungal diseases.

THE ORGANIC VINEYARD

An organic vineyard is one that is grown without the use of any synthetic fertilizers, weed killers, insecticides, or other chemically manufactured products. To be able to sell your grapes as organic, the vineyard must be certified by an organic organization such as the National Organic Program (NOP), the National Sustainable Agriculture Information Service (ATTRA), or the Canada Organic Regime. In addition to these federal organizations, there are organic and sustainable farming organizations at the state and provincial levels.

Sustainability is a process of growing crops using practices that do not use up the natural resources of the land, but rather enhance and improve them. A sustainable farmer or grower aims to prevent soil depletion and erosion, water pollution, loss of biodiversity, and ecological impacts from crops. That means improving the texture and fertility of the soil by planting cover crops and using amendments such as compost and manure. Above all, the sustainable grower encourages resilience in the crops rather than chemical dependence, and uses the resources at hand to help maintain good health in the soil, water, and crops.

Managing Problems Organically

The concept of Integrated Pest Management (IPM) was developed by the University of California, Davis. This system follows the principle of employing the least toxic pest control practices. Instead, it aims to harness existing methods of control whenever possible. For instance, one way to encourage natural pest control is to maintain a nearby hedgerow of native and other plants that attract beneficial predator insects like wasps and spiders. These insects in turn feed on vineyard pests such as mites and thrips. I have such a hedgerow alongside my own vineyard. In addition to using such principles, the sustainable grower will choose organic pest and disease control products that can address the most common problems faced by grapevines. Spraying basic elements like sulfur and copper to combat fungal diseases is one example.

The greatest concern of an organic vineyard is that a pest or disease may invade that is not treatable by anything other than a chemical compound. But this is becoming increasingly rare as more organic treatments are developed. In my experience, this has not been an issue.

I believe you can have an organic vineyard almost anywhere you can have a vineyard. If you already have a conventional vineyard, you can convert it to a more organic and sustainable vineyard by eliminating chemicals and improving your soil health and vineyard biodiversity. Under the National Organic Program, you may even be certified over a three-year transition process.

Of course, you do not need to be officially certified in order to grow your own grapes for winemaking. But an organic and sustainable vineyard is still a good choice for home growers and winemakers for a variety of reasons. One is that handling chemical pesticides often requires permits and extra precautions, by anyone who is working in the vineyard and even by your neighbors. These products also cost money; an organic vineyard is generally less expensive to run than a conventional one (although it may take more labor). Finally, you can contribute your part to being less of an impact on your community and the world.

Winegrowing Around the World

Two broad bands that circle the globe are suitable for growing grapes. These temperate regions have sufficient summer high temperatures, winter low temperatures, and a growing season that lasts long enough to ripen grapes. Within these bands are a wide variety of regional differences.

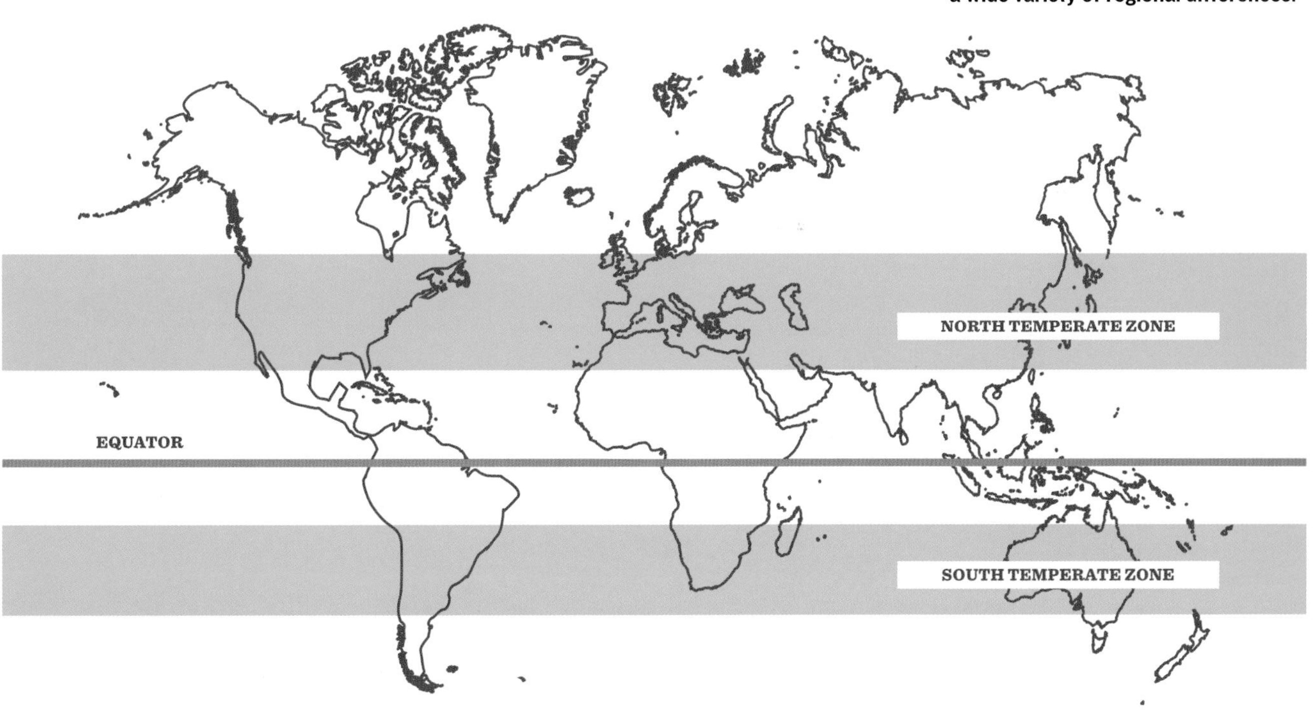

Thomas Jefferson was one of the earliest vintners in the United States, experimenting with both European and American grapevines. Today, Virginia wineries continue to make wines with vinifera grapes as well as local varieties like Norton.

YOUR LOCATION

Let's begin our discussion of creating your vineyard with location. You probably have a pretty good idea of whether your macroclimate is one in which grapes can grow. If there are already vineyards in your region, then you know you're in a suitable macroclimate. Basically, you need at least eight hours of sun per day during the growing season. In the Northern Hemisphere, this is from about February to April through September to November. This covers a great deal of the United States and parts of southern Canada.

Of course California has the greatest abundance of different growing regions in North America. But there are significant grape-growing areas in the Northeast—in New York state, southern New Jersey, Pennsylvania, and parts of Ohio. Some of the original vineyards in the United States were planted in Virginia, and there are vineyards in Maryland and North Carolina. In the South and Southeast, grapes are grown in Kentucky, Arkansas, Oklahoma, Florida, Texas, and New Mexico. In the Midwest, a growing body of knowledge has developed for cold-winter growers in Michigan, Minnesota, Missouri, Illinois, Indiana, and Wisconsin. The Northwest has many fine commercial and hobby vineyards in Oregon, Washington, and British Columbia. And

in eastern Canada, there are thriving wineries in Ontario, Quebec, and the Maritimes. In fact, there are vineyards to be found in some of the most extreme locations of cold, heat, and humidity throughout North America.

Local Conditions

The next consideration is your mesoclimate—the conditions in your specific local area. What are the differences and similarities within your regional climate where grapes are grown and the location of your planned vineyard? Are you on the side of a mountain or beside a lake? Does your property slope? If so, in which direction? Are there other weather-moderating factors nearby?

Some differences are probably not of great importance in climatic regions that have average summer temperatures, say 60°F to 85°F (15°C to 30°C), but they are important where the macroclimate conditions are marginal for vineyards. In these areas, differences in mesoclimates can be measured in distances from a few yards up to many miles.

In cooler areas, you can increase your chances of more sun by planting on a southern slope. Likewise, in marginally cooler areas where chances of frost exist during the growing season, exposure of the vineyard to a lake or bay will moderate the potential for early or late frosts.

Cold winters are no deterrent if you wish to plant a vineyard. In Ontario's Niagara region, grapevines overwinter with no ill effects. The region favors cool-climate varieties like Pinot Noir and Riesling.

The Texas hill country is among the fastest growing wine-producing areas in the United States. The hot summers are suited to varieties that love plenty of heat and sunshine, just like the native bluebonnets and other wildflowers that blanket the hills.

In hotter areas, if you are on a hill, where breezes can cool the high vineyard temperatures within plants' preferred temperature range—not more than 95°F (35°C constant)—your vines will do better. Grapes cannot thrive at altitudes above about 3000 feet, however. The cooler temperatures and the swings between daytime and nighttime temperatures do not favor the development of good wine grapes.

WHAT VINEYARDS NEED

- Locate the vineyard so that during the main part of the growing season it gets at least eight hours per day of sunlight.
- Be sure there are no obstructions to the sun from houses, fences, garages, barns, or other man-made structures, or from trees, shrubs, bushes, or other vegetation near the vineyard.
- Be sure there are no obstructions from within the vineyard such as rows spaced too closely together so that they shade one another. Do not space rows closer than five feet apart, except in special circumstances.

Sandy soils and the maritime climate on New York's Long Island have inspired both commercial and home winemakers to plant vineyards in fields that were once potato farms.

Obstacles That Shade the Vineyard

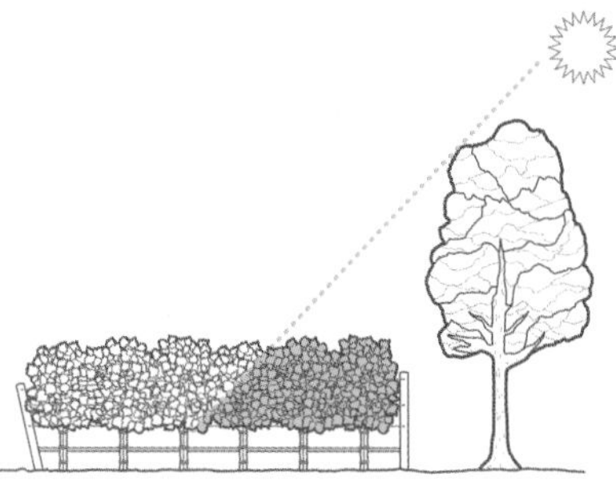

Trees cast shade. Remember that small trees at the edge of the vineyard will grow in the future.

Vines planted too close to a house or other building may be too shaded to produce good fruit.

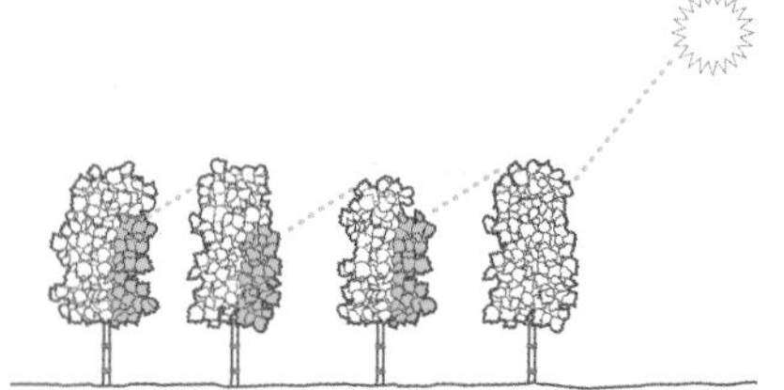

Row spacing is important to ensure that the vines do not shade those in adjacent rows.

- Know your azimuth! The direction the sun moves over your vineyard during the growing season will allow you to determine how to orient the rows in your vineyard. The further north you are, the more important this becomes. Throughout North America, south-facing slopes will receive more direct sunlight than any other exposure. The next best are west-facing slopes, which receive afternoon sun. East-facing slopes get more sun in the morning. One exception here is that in very high-temperature areas you may want to run rows east-west to prevent sun burning the grapes, particularly in the afternoon.
- Do not plant a vineyard in a spot that is exposed to severe winds. Strong or steady winds can desiccate plants and cause soil erosion. If necessary, install windbreaks to moderate the effects of wind; this may be a tall hedge, a fence, or trees. Windbreaks should not be solid, but allow some air to pass through.
- Slopes are not necessarily bad for vineyards, but grapevines planted on a slope can be more difficult to care for, especially with mechanical equipment such as tractors or mowers. Also, vines planted lower down on the slope can be subject to colder air that naturally flows down into a hollow. You may also need to consider erosion control. If your property slopes steeply, get advice from a vineyard consultant before planting.
- Water is an essential element for a successful vineyard, particularly in arid climates. If you have a backyard, chances are you have convenient access to water. One thing about vines is that they need much less water than traditional landscaping. If you are used to a suburban yard with a lawn and landscaping, you may be tempted to water the vineyard more than necessary. This is unwise, as too much water in the vineyard can cause as many problems as too little water.

Tips for Late-Frost Climates

Frost occurs when the temperature drops below freezing (32°F or 0°C). Frost damages young growth and buds so if you live in an area that receives occasional late frosts, you can reduce the potential for frost damage in an orchard or vineyard with the practices outlined here. If you do not know your average frost dates, contact your local extension service or your local Environment Canada office.

- Frost-sensitive crops like grapes should not be grown in low areas where cold air is trapped by natural topography, vegetation, or by manufactured obstacles like fences that can block the drainage of cold air. Select your site carefully.

- Plant on south-facing slopes. Always try to plant on a hillside if you can, as it is the valleys where frost hits the hardest. Hillsides usually have just enough wind to prevent the frost from settling.
- Select grape varieties that develop later in the spring, so that the developing buds aren't damaged by late frosts. Swelling buds are not sensitive to frost, but any exposed green growth can be damaged. Varieties with later bud break will survive frost, but you must choose varieties that have enough time to ripen in your growing season.
- Manage soil and ground cover plantings to maximize the storage and later release of heat from the soil within and upslope of a crop. These passive management practices are less expensive than the active frost protection practiced by commercial vineyards, which may involve sprinkler irrigation and wind machines.
- To reduce the chances of frost damage, make sure that the ground is firm, moist, and exposed to sunlight by:
  - eliminating or cutting ground cover weeds and planting cover crops
  - keeping the top 1 foot of soil moist
  - not cultivating until all danger of frost has passed

In cold winter regions in Canada, icewine has become a specialty, and it's possible to make it in a home vineyard. Choose a white grape like Riesling or Vidal and let the grapes hang on the vine until they are frozen. Netting protects them from hungry birds.

- Prune as late in the dormant season as possible.
- One specialty that is possible in colder climates is icewine. This wine is made by allowing ripe wine grapes to freeze naturally on the vine. The grapes then are picked by hand when temperatures are 9°F to 18°F (–8° to –13°C). This allows the correct level of sugar and flavor. Do not expect to produce a lot of wine, perhaps only a bottle per vine, but the wine is very sweet and golden rich.

Tips for Hot-Summer Climates

In areas where summer temperatures often reach above 95°F (35°C), such as inland deserts in California and the Southwest, I would consider the following:

- Plant varieties that tolerate heat and still develop well-balanced wine, such as reds like Syrah, Tempranillo, and Verdelho, and whites like Viognier, Sauvignon Blanc, and Chenin Blanc.
- Orient the rows east-west so that the sun passes over the top of the vines.
- Don't pull the leaves on the 'afternoon' side of your rows. Too much leaf pulling can cause sunburn.
- Consider alternatives to the VSP trellis. There are some trellis designs that promote shade.
- Irrigate more near harvest time to slow the ripening of the grapes.

Tips for Short-Summer Climates

In cooler climates that have a shorter growing season and many overcast days, you will need to concentrate on getting fruit to ripen. This is an issue for vineyards in places like the Pacific Northwest, where late summer rains, fog, and cloudy days are common.

- Choose the warmest possible spot, such as a south-facing slope for your vineyard.
- Orient the rows north-south to maximize the amount of sunlight.
- Choose varieties that can ripen with a shorter growing season, such as Gewürztraminer, Madeleine Sylvaner, New York Muscat, Pinot Blanc, and Pinot Noir.
- Spread black mulching material or rocks under the vines to absorb heat.
- Pull leaves to expose as much of the canopy to sunlight as possible.
- Develop a good program of control for fungal diseases, which are often worse in cool, damp climates.

TEMPERATURE AND VINE GROWTH

Grapevines are deciduous, woody plants. This means that they develop a hard, woody permanent framework that puts out new growth during the warmer months and then enters a period of dormancy during the winter months. It helps to know the stages of growth because timing of various tasks is extremely important when maintaining your vineyard.

Dormancy is the period when your vines are growing very little or not at all. Depending on the soil temperature, the vine may still have some root growth, but where the soil is frozen, there is likely to be very little activity even in the root zone. Grapevines need this period of cold, which is why they do not grow in tropical climates where there is no winter freeze. Some grapevines are able to withstand winter temperatures as low as −45°F (−42°C). The degree of cold-hardiness depends on the variety and the rootstock.

In spring, rising temperatures stimulate the vine to start converting the starch it has stored in its woody tissue to sugar. Sugary sap moves through the vine, and growth of the green tissues starts.

Bud break or bud burst occurs when the average daily temperature is 50°F (10°C). Shoot development begins right after bud break (the leaves and fruit are produced from these shoots). Once leaves grow to about half their final size and lose their initial bright green color, they begin to manufacture new sugar by photosynthesis.

Flowering occurs 30 to 80 days after bud break at 61°F to 68°F (16°C to 20°C). The inflorescences begin to swell, then open to reveal the small flowers. Most grapevines are self-fertile, which means they do not need to have male and female plants in order to pollinate. Wind and insects do assist with pollination, however.

Berry development begins with berry set or fruit set, which comes 7 to 10 days after flowering and ends with harvest. In hot climates, the total period from bud burst to harvest is shorter (110 to 140 days); in cooler climates, it is much longer (190 to 220 days).

Veraison is the mid-stage of berry development when the berry softens and changes from green to its final color (amber, black, blue, golden, pink, red, or yellow).

After harvest, the green leaves continue to photosynthesize. Leaf fall is stimulated by frost or water stress. After leaf fall, the vine is dormant during the winter months. Sugars are converted to starch and stored in the trunk and roots.

FROM TOP:
Just before bud burst, the developing nodes or buds push out from the cane in response to rising temperatures. | After bud burst, you will see the developing shoots. | The grapes emerge near the base of the canes, first as small clusters of buds. | The flowers are visible as the clusters start to open up. | As the fruit develops and the clusters get heavier, they will start to hang downward.

ASSESSING THE SOIL

In Paris in 1973, a blind tasting was held with professional wine tasters from France. The purpose was to compare wines from France with those from the Napa Valley in California. It was inspired by the California vintner Robert Mondavi, and resulted in the American wines being chosen as winners over many of the French wines.

Needless to say, this was a big surprise to the French, who had heretofore said that the best premium quality grapes could only be grown in French soils. How wrong they were. Soil type is important, but it is by no means the most important factor. In fact, studies since the 1970s have shown that premium wine grapes can be grown in a variety of soils.

Most topsoils are categorized as clay, loam, or sand. In fact, many soils are a mixture of two of these types. Clay soils are typically heavy, do not drain well, and are nutrient-rich. This type of soil holds water and is poorly aerated, which limits root depth. Silt is a fine type of clay soil. Sandy soils drain well and tend not to retain nutrients, but they are well aerated and allow roots to penetrate deeply and easily. Loam or loamy-textured soils are mixtures of particles of different sizes, a blend of sand, silt, and clay. Loam soils usually are rated highest in agriculture because they have the best mix of water, air, minerals, and nutrients for plant growth. They also contain plenty of organic matter, which supports biodiversity in the soil and is essential for soil health.

Grapevines are adaptable to most of these soil types, but like all plants, they grow best when they have healthy and well-developed root systems. Grapevine roots can grow down as much as 15 feet in deep soils, but the main feeding roots are usually concentrated in the top 2 or 3 feet of soil. They cannot develop these root systems without soil of adequate quality. Shallow soils may be limited by boulders, bedrock, hardpan, a high water table, or some kind of mineral or chemical barrier.

Still, many people worry too much about the kind of soil they have. In general, there will be a rootstock that will adapt your soil to the grape variety you select to grow. That is not to say that you can put your vines into the soil without a second thought. Soil amendments, quality of sunlight, drainage, water supply, and other issues can moderate differences in soils. To have a healthy and well-developed root system, it makes sense to be aware of the characteristics of good soil and how to deal with poor soil.

Soil Qualities

The best soils:

- Are least 3 feet deep, although you can grow vines in less depth. If the soil is rich you can use a rootstock that spreads out laterally rather than growing deeply down into the soil. This will normally require you to provide wider spacing between plants and rows. You will also need to choose varieties that can adapt to shallow rootstocks.
- Have low mechanical resistance (no big rocks, hard uncultivated layers, or other such barriers).
- Are well aerated (contain plenty of oxygen) and well drained (don't have standing water during the growing season).
- Have optimum temperatures, and pH levels between 5.5 and 7 pH.
- Contain no toxic substances or contaminants.

The first three characteristics on this list can probably be determined by a simple judgment call on your part. If you are not confident in the quality of your soil, then dig a hole in the proposed site for a better evaluation. The hole should be at least

A vineyard planted on a hill can be more challenging in terms of equipment and labor, but in marginal climates a south-facing slope will maximize the amount of heat and sunlight that the grapes receive.

3 feet deep, so you may wish to rent a post-hole digger. After you dig the hole, fill it with water and evaluate how long it takes for the water to drain. If the hole is still full of water after 24 hours, then you may have a drainage problem.

To improve drainage in soils, the first option is to till it (loosen and aerate). You can do this with a rototiller, a discer, or a backhoe, as long as you break up the earth down at least a foot. If the soil is very compacted, you may want to add some soil amendments when you are tilling. In some cases, you may have to remove big rocks and other barriers and improve drainage using subterranean drains (such as French drains) to remove midyear standing water. Before embarking on an underground drainage system, speak to a vineyard consultant or a soil engineer.

Optimum temperatures, pH, and the presence of contaminants are harder to judge without soil testing. If you are planting a small home vineyard and you do not want to do soil tests, first take a look at the property. Do other plants such as grasses, shrubs, or flowers thrive there now? Barren ground, similar to that typically found under pines or other trees, is a good clue you do not have a well-balanced soil, or you have compacted soil.

Soil Nutrition

There are three key nutrients and nine other micronutrients in soil that have an impact on grape vine and wine quality. These vary from one region to another, and even from one backyard to another.

The best vineyard soils combine fertility and good drainage. Too many obstacles, such as stones and boulders, can obstruct good root growth. Try to remove as many big rocks as is practical.

MAJOR NUTRIENTS

- Nitrogen (N) ensures overall structure and metabolism; promotes green growth.
- Phosphorus (P) promotes flowering and strong structures (roots and trunk).
- Potassium (K) promotes flowering, fruit production, and general health.

Micronutrients

- Boron ensures sugar and generalized metabolic movement.
- Calcium supports leaf and root growth. Acid soils (low pH) may be low in calcium. So-called calcareous soils are high in calcium carbonate and are alkaline (high pH).
- Copper supports chlorophyll synthesis, leaf growth, and lignin formation.
- Iron promotes chlorophyll formation. Lack of iron can lead to chlorosis, which is evidenced by lack of color in the leaves.
- Molybdenum helps in conversion of nitrates and nitrogen-fixing bacteria in the soil.
- Magnesium supports chlorophyll formation. Deficiencies sometimes found in sandy soils.
- Manganese is involved in synthesis of chlorophyll and metabolism of nitrogen.
- Sulfur promotes amino acid and chlorophyll formation. Lack of sulfur can also show as chlorosis.
- Zinc supports flowering and fruit set. Lack of zinc can also lead to small leaves ("little leaf disease").

Soil type and structure do have an impact on grape vine nutrition and management of these nutrients. Also, the pH of the soil has an influence on nutrients in the soil and the environment around the roots. pH levels that are too high or too low can restrict the plants' ability to absorb nutrients, even if those nutrients are present in the soil. That's why a soil test for agricultural values and deficiencies is preferable before planting.

Your local extension service or agriculture office may provide soil testing, or you can use a commercial service. Complete instructions for the soil test will be provided by the testing service. Normally you will have to take a number of samples from different areas of the proposed vineyard.

One of the best ways to improve soil fertility and texture is to maintain a good compost pile filled with clippings from your vineyard and garden, and with organic scraps from the kitchen.

Soil Testing

A typical report will tell you:

- Your soil type: clay, loam, or sand.
- Soil pH, which indicates if the soil is acidic or alkaline. Most grapevines prefer a pH level between 5.5 and 7 pH.
- The presence of macronutrients (nitrogen, phosphorus, potassium), micronutrients (boron, calcium, copper, iron, molybdenum, magnesium, manganese, sulfur, zinc), and salts.
- The amount of organic matter in the soil.
- Pathogens, or soil-dwelling pests such as nematodes or phylloxera.
- Recommendations for soil amendments to correct nutrient deficiencies and pH.

Maintaining cover crops and ground covers between the vineyard rows can add fertility to the soil, prevent weeds, and attract beneficial insects.

2

CHOOSING YOUR VINES *and* ROOTSTOCKS

POPULAR GRAPE VARIETIES

PLANNING YOUR VINEYARD

THE TRELLIS SYSTEM

VINEYARD IRRIGATION

CHOOSING YOUR **VINES AND ROOTSTOCKS**

All grapevines belong to the botanical genus *Vitis.* The traditional species of grapevines grown for use in winemaking is the European species *Vitis vinifera* (often called simply "vinifera"), whose origin is in and around the Mediterranean, including Europe. The best known American grape species is *V. labrusca,* native to eastern North America from Ontario as far south as Louisiana. When talking about grapes and wine, we use the variety name (such as Chardonnay or Pinot Noir) unless we are talking about a species or a hybrid cross between two species. American varieties and hybrids developed from North American parents typically require less summer heat and tolerate greater winter cold than vinifera grapes. Generally speaking, vinifera grapes grow well in the West and Northwest, and in some other areas with the right conditions to ripen the grapes. Typically, colder areas in the Northeast and Midwest are best suited for American varieties and American hybrids. In the humid conditions of the South, native muscadine grapes may be the best choice.

We always refer to two different parts of the plant: the rootstock and the scion or clone. The rootstock is the part of the plant below and just above ground, including the roots and the bottom part of the trunk. Grafted onto the top of the rootstock is the

Base of Vine

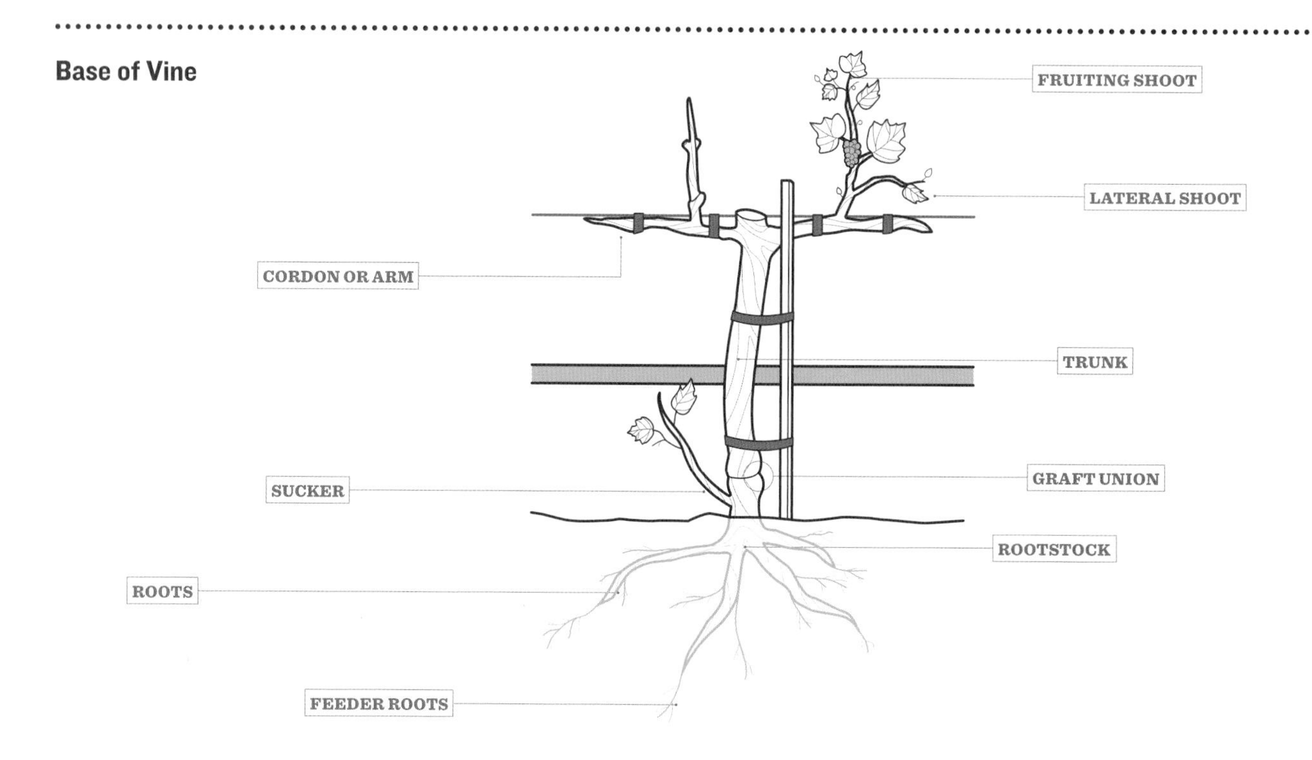

scion or clone, which is a variety such as a Cabernet or Chardonnay. This includes the upper part of the main trunk, the branches, leaves, flowers, and fruit. The term scion refers generally to that part of the plant above the rootstock, whereas the clone is a variety and subspecies, such as Cabernet Sauvignon 337 or Merlot 182. Some varieties have many different clones that are suited to different growing conditions.

Buying Grapevines

Some history is important here. The development of rootstocks and varieties goes back to antiquity. Most of the early definitive work was done by German, French, and Italian grape breeders, but in the late 19th century, the University of California, Davis, and Cornell University in New York state were also researching grape breeding. And, particularly since the 1970s and 1980s, breeders have also developed many rootstocks and varieties that are suited to North American soils and climates.

For each region, and even each vineyard, there are some choices of both rootstock and grape variety that are better than others. The simplest way to choose grape varieties is based on your taste, and by finding out which grapes are being successfully produced in your local area. However, I am willing to stick out my neck somewhat and make a few recommendations based on my experience and knowledge.

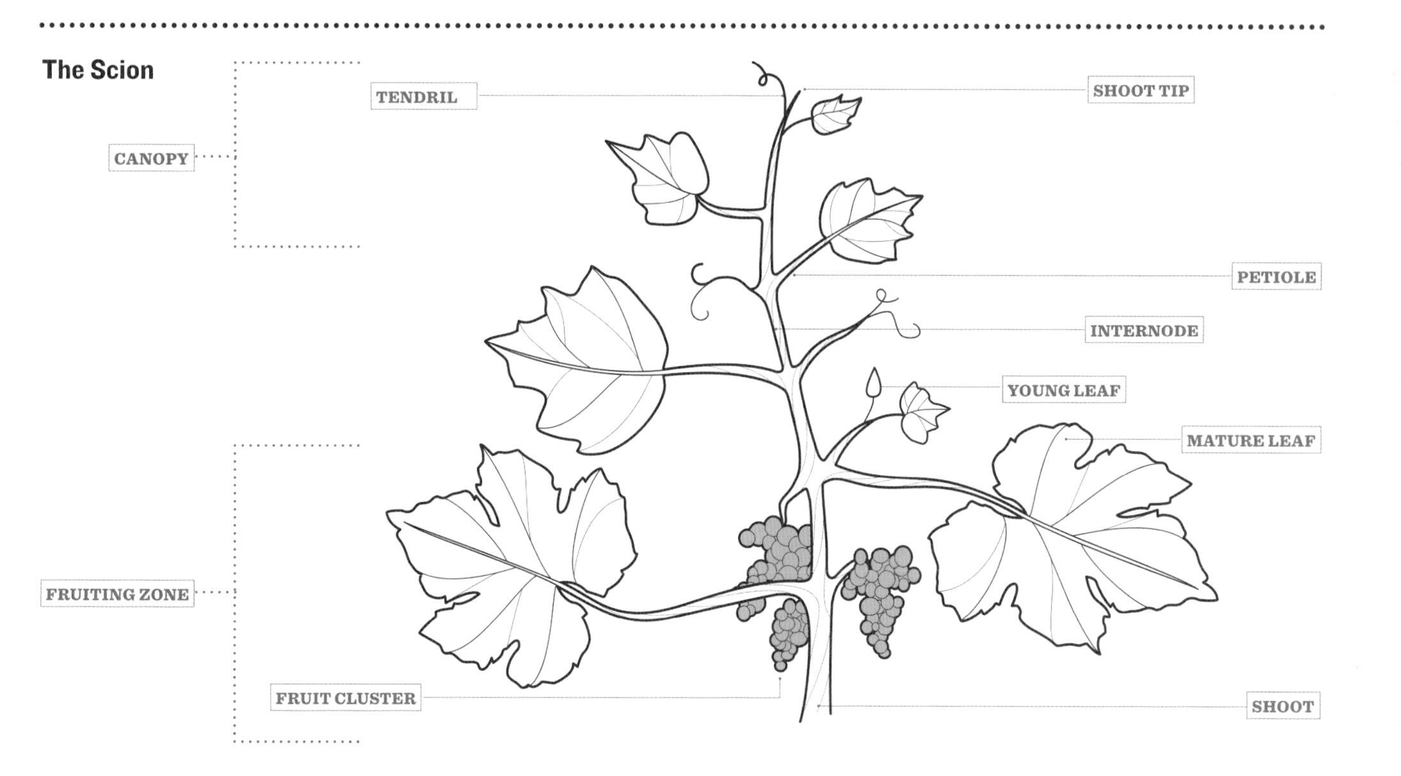

GRAPE VARIETIES BY AVERAGE DAILY TEMPERATURE

60°F to 85°F (15°C to 30°C)
The grapes that can be grown in climates that are in the midrange of temperatures are those you've probably heard about in the premium wine trade, such as Cabernet Sauvignon, Chardonnay, Merlot, and Pinot Noir, to name a few.

85°F to 95°F (30°C to 35°C)
In these upper temperature ranges, you will need heat-loving varieties such as Syrah, Colombard, Ruby Cabernet, and Grenache.

50°F to 60°F (10°C to 15°C)
These are challenging temperatures, but it's not impossible to grow good grapes varieties such as Gewürztraminer, Riesling, and Pinot Noir.

I will also suggest what I believe are some of the best rootstocks for home vineyards.

Different nurseries will carry different clones, depending on availability and demand. I strongly recommend that you buy your grapes from a reputable supplier, whether locally or by mail order. For recommendations, ask your county, state, or other local agricultural advisor, or get advice from nearby growers, a Master Gardener group, or a local vineyard supplier. Most important is that the nursery should sell certified plant stock. The California Department of Food and Agriculture has a certification program, as do many states and provinces with active vineyard communities.

Ordering from out-of-state or out-of-country can sometimes be difficult because of quarantines, but if you find a good out-of-state source, check with your extension service or agriculture office for the correct procedures for importation.

Matching the Grape to your Climate

Many places throughout the world that have not been historically noted for grape growing have discovered that there are grapes that can grow successfully if the mesoclimate is identified, the vineyard is ideally located and constructed, and the proper grape variety is selected.

A key thing to know about varieties is the number of frost-free days they need to ripen; that is, the period between bud burst and harvest. Some varieties need as few as 110 to 140 growing days to ripen. Most vinifera grapes need between 150 and 200 days. If you live in an area with short summers, you must choose varieties that will have time to ripen before temperatures start to drop in fall. Keep in mind that even if you have potentially enough frost-free days, if your vineyard is typically subject to heavy rainfall toward the end of the growing season, you should choose varieties that can ripen before their onset.

POPULAR GRAPE VARIETIES

	WHITE	COMMENTS
	Aurora	Hybrid developed in the 1860s. Very early; highly productive. Susceptible to splitting fruit and black rot. Best in East and Midwest. Hardy to –15°F (–26°C).
	Aurore	French-American hybrid. Early ripening. Best for cooler areas. Hardy to –15°F (–26°C). Makes a light, mild wine. Susceptible to bunch rot, powdery mildew, black rot.
	Cayuga	Hybrid developed at Cornell in New York state. Productive. Hardy to –10°F (–23°C). Midseason. Disease-resistant. Makes good quality wine.
	Chardonnay	Classic white burgundy grape. Many clones available for different regions. Early bud break, so susceptible to late spring frosts. Hardy to 0°F (–17°C). Susceptible to Pierce's disease.
	Chenin Blanc	Susceptible to Pierce's disease, bunch rot. High-yielding vines make a simple white wine. Hardy to –5°F (–20°C).
	Edelweiss	One of the hardiest and most disease-resistant white hybrids. Bred from *Vitis labrusca* parentage. Hardy to –25°F (–31°C). Golden fruit makes a sweet white wine.
	Ehrenfelser	A cross of Riesling and Sylvaner, a German hybrid widely grown in British Columbia.

Chardonnay

Frontenac Gris

Gewürztraminer

POPULAR GRAPE VARIETIES *continued*

WHITE	COMMENTS
Frontenac Gris	Hybrid developed at University of Minnesota. Hardy to −30°F (−35°C). Midseason ripening.
Gewürztraminer	Midseason ripening; hardy to −10°F (−23°C). Best in cooler regions. Makes a spicy, fruity, or dessert wine.
Malvasia	Good white for hot areas. Dry, light-bodied wine. Hardy to 0°F (−17°C).
Marsanne	Traditional Rhone variety; makes a full-bodied white. Good for hot areas. Hardy to 0°F (−17°C).
Niagara	Hardy to −15°F (−26°C); ripens late midseason. An American white widely planted in New York state; moderately disease-resistant. Variety of *Vitis labrusca.*
Orange Muscat	Intensely flavored muscat grape; hardy to −10°F (−23°C).
Ortega	Very productive hybrid; makes a light, fruity wine. Popular on Vancouver Island, BC. Can be used in icewine.
Pinot Blanc	Best in cooler parts of hot climates. Makes a light white wine that can be barrel aged.

POPULAR GRAPE VARIETIES *continued*

WHITE	COMMENTS
Pinot Gris (Pinot Grigio)	Hardy to –10°F (–23°C); ripens midseason. Not highly productive, but good choice for a light, fruity wine.
Ravat 51	Hybrid grown in cooler climates; makes a fruity, semi-sweet wine. Can be used for icewine or late-harvest wine. Hardy to –25°F (–31°C).
Riesling	Hardy to –5°F (–20°F); late-ripening. Susceptible to cluster rots at harvest. Makes wine with characteristic German-style bouquet.
Roussanne	Rhone variety often grown with Marsanne and Viognier. Good for hot areas. Hardy to 0°F (–17°C).
Sauvignon Blanc	A Bordeaux white. Many clones available for different regions but likes plenty of sun. Susceptible to Pierce's disease, bunch rot.
Semillion	Vigorous white Bordeaux grape; prefers cooler climates such as the Northwest. Hardy to –10°F (–23°C).
Seyval	Hybrid hardy to –15°F (–26°C). Early bud break and ripening, susceptible to powdery mildew and bunch rot. Most popular wine grape in the eastern U.S.
Verdelet	Late-ripening French-American hybrid. Hardy to –10°F (–23°C). Vigorous; may need thinning. Can be used for icewine.

Pinot Gris *Riesling*

POPULAR GRAPE VARIETIES *continued*

	WHITE	COMMENTS
	Vidal	Late-ripening French-American hybrid. Hardy to –10°F (–23°C). Vigorous; may need thinning. Can be used for icewine.
	Villard	Hybrid; makes a good, light wine. Fairly long growing season. Hardy to –10°F (–23°C).
	Viognier	Along with Roussane and Marsanne, one of the Rhone varieties. Makes a rich, flavorful wine. Can be prone to diseases. Hardy to 0°F (–17°C).
	RED	**COMMENTS**
	America	Hybrid hardy to –30°F (35°C); tolerates Pierce's disease. Ripens midseason.
	Baco Noir	Hybrid hardy to –20°F (–29°C); very vigorous and productive. Ripens early. Susceptible to black rot. Good for heavy soils.
	Barbera	Popular Italian variety. Vigorous but susceptible to disease. Hardy to 0°F (–17°C).
	Black Spanish ('Lenoir')	Hybrid for hot climates; tolerates Pierce's disease. Ripens midseason. Used to make Texas port.
	Cabernet Franc	Similar to Cabernet Sauvignon but more hardy, to –10°F (–23°C). Good for cool-climate East Coast vineyards.

Frontenac

Gamay

Merlot

Pinot Noir

POPULAR GRAPE VARIETIES *continued*

RED	COMMENTS
Cabernet Sauvignon	Bordeaux variety that ripens late and is not hardy below 0°F. Susceptible to diseases, including powdery mildew and downy mildew.
Chambourcin	Moderately vigorous hybrid hardy to –10°F (–23°C). Resistant to powdery mildew. Good for lower Midwest and East Coast.
Chancellor	Hybrid hardy to –20°F (–29°C); early bud break. Very productive; may need thinning. Susceptible to downy mildew.
Chelois	Hybrid primarily used as a blending wine. Susceptible to bunch rot; hardy to –15°F (–26°C).
Concord	*Vitis labrusca* variety. Widely grown American red; good disease resistance. Characteristic slip skin and foxy flavor. Hardy to –25°F (–31°C).
de Chaunac	Very productive and dependable hybrid. Late bud burst; ripens midseason. Hardy to –15°F (–26°C).
Delaware	Used for table grapes and wine. Ripens early and likes a deep, fertile soil. American red widely planted in the Northeast. Hardy to –25°F (–31°C).
Dornfelder	German red wine grape hardy to –10°F (–23°C). Highly disease-susceptible.

POPULAR GRAPE VARIETIES

RED	COMMENTS
Foch	Early ripening hybrid hardy to −25°F (−31°C). Productive; makes a good blending wine.
Frontenac	Hybrid developed at the University of Minnesota; hardy to −35°F (−37°C). Moderate to late bud break; midseason ripening. Susceptible to powdery mildew. High sugar in cooler climates.
Gamay	Vigorous Beaujolais variety. Not hardy below 0°F (17°C); best in hot climates.
Grenache	More widely adapted than many reds. Very vigorous, drought- and heat-tolerant. Not hardy.
Leon Millot	Hybrid hardy to −15°F (−26°C), vigorous, and disease-resistant.
Malbec	Bordeaux red that makes full-bodied wine but usually blended. Sensitive to downy mildew. Needs sun and heat but planted in some marginal areas.
Marechal Foch	Early bud break and ripening. Productive hybrid. Hardy to −25°F (−31°C); well suited to the Northwest. Needs winter protection in coldest climates. Also called Foch.
Merlot	Makes a soft, full-bodied red. Hardy to 0°F (−17°C); disease-susceptible.

POPULAR GRAPE VARIETIES *continued*

	RED	COMMENTS
	Mourvedre	A Rhone red that likes hot weather; not very hardy, to 0°F (–17°C).
	Muscadine (*Muscadinia rotundifolia*)	Native to the southeastern U.S.; best in hot, humid climates. Many varieties, including Scuppernong, Black Beauty, and Summit.
	Nebbiolo	Heat-loving Italian variety. Not for cold regions.
	New York Muscat	Blue grape hybrid that ripens early. Hardy to –15°F (–26°C). Best for cooler climates.
	Norton (*Cynthiana*)	Blooms late; likes good drainage. Hybrid widely grown in the Midwest, North, and Northeast; hardy to –20°F (–29°C).
	Petit Sirah	Grown mostly in California; produces full-bodied wine. Hardy to 0°F (–17°C). Susceptible to disease.
	Pinot Noir	Many clones available for different areas. Very adaptable grape. Ripens mid-season; hardy to 0°F (–17°C).
	Rougeon	Hybrid grown in the Finger Lakes region; hardy to –15°F (–26°C). Often used for blending.

POPULAR GRAPE VARIETIES

	RED	COMMENTS
	Ruby Cabernet	Cross between Cabernet Sauvignon and Carignan developed for hot areas, especially in California.
	Sangiovese	Italian variety. Susceptible to disease. Early bud break; hardy to 0°F (–17°C).
	Schuyler	Very productive hybrid, hardy to –15°F (–26°C). Ripens early to midseason.
	Syrah	Loves heat but can also tolerate cooler climates. Big flavors in the wine, heavy crop and vigorous growth. Hardy to 0°F (–17°C).
	Tempranillo	Spanish grape for hot climates. Hardy only to 0°F (–17°C).
	Zinfandel	One of the grapes originally widely dry-farmed in California. Susceptible to disease, vigorous. Hardy to 0°F (–17°C).

CHOOSING **ROOTSTOCKS**

Why do you have to choose a different rootstock for your grapevines rather than let them grow on their own roots?

The first reason is that certain rootstocks are bred to be disease- and pest-resistant. At one time, most grapevines were grown on their own roots. But after the American grapevine pest phylloxera was unintentionally exported to Europe and destroyed many vineyards planted with own-root vines, grape growers began to graft varieties onto American rootstocks that are resistant to phylloxera. There are rootstocks that are also resistant to other common vineyard pests, such as nematodes.

Secondly, different rootstocks are adapted to different soil and weather conditions. This means that you can choose the rootstock that best suits the conditions you have and the qualities you are looking for.

Finally, the vigor of a rootstock must be matched to your type of soil and conditions. A vigorous rootstock planted in rich soil may produce more growth that you can handle, and you will find yourself constantly pulling off leaves and snipping off young clusters of grapes. If you do plant a vigorous grapevine, consider a wider spacing, such as 7 or 8 feet. Similarly, vines with less vigor may need to be planted with a closer spacing to achieve sufficient fruit yields.

ROOTSTOCK QUALITIES

- **Vigor:** Encourages the vine to produce more or less vegetative growth (leaves) and fruit.
- **Hardiness:** Makes the vine able to withstand a greater amount of winter cold.
- **Ripening:** Reduces or lengthens the number of days needed to ripen the fruit.
- **Soil adaptation:** Allows a vine to grow in soils with particular characteristics, such as extremes of pH (alkalinity or acidity).
- **Diseases and Pests:** Makes a variety resistant to certain soil-borne diseases or insects, especially nematodes and phylloxera.

You can get a wide selection of rootstocks and varieties in several different forms. Grafting is the process of joining the scion to the rootstock, which may be done either in the field (field-grafted) or in the greenhouse (bench-grafted). I am going to suggest you get either a green-growing or dormant bench-grafted vine. I have had very good experience with these kinds of vines as have many other growers, both commercial and amateur.

Nurseries tend to have potted 3-inch green-growing plants from early spring into the summer. These are young vines that are in active growth, with several green shoots and root systems with plenty of new growth, similar to a container-grown shrub or vine you would purchase at the garden center. If you are planting during the growing season and can get some of these green-growing plants, that's what you should do. During late fall or winter you will typically get dormant vines from the nursery. These are the easiest to get if you source your vines from far away, as it is easier to ship dormant vines in plastic with damp sawdust. They are also cheaper than green-growing vines. The beginning grower often worries about dormant vines because they look dead when they arrive. But that is only because dormant vines are not in active growth and they have been kept in cold storage before shipping. You can still see a developed root system on the plant, and they will quickly send out new roots and green growth.

Recommended Rootstocks

To make things simple, I am going to limit the number of rootstocks to some tried-and-true selections. However, you may want to consult on your own with a qualified local grower, your extension service, or a nursery. When you first start to look at possible varieties and rootstocks, the lists can seem very long. I have found nursery staff to be generally very helpful in advising you about what will work for your conditions.

- **110 R:** Vines grafted on this rootstock are vigorous in fertile soil, ripen their grapes well, and have high sugar content. This rootstock accommodates well to all kinds of soils, and it is excellent in warm grape-growing areas with an arid climate. It is phylloxera-resistant and moderately resistant to nematodes, and it tolerates up to 17 percent active lime in the soil, commonly referred to as acidic soil.
- **1103 P:** This rootstock grafts and roots readily, and is compatible with most clonal varieties. It is adaptable to calcareous clay soils with fertile subsoils, and accommodates better to very dry conditions (warmer climates) than 110 R. It is phylloxera-resistant and moderately resistant to nematodes.
- **5 BB:** This is a very important rootstock in areas with shorter growing seasons, as it has a shorter growing cycle. This is classified as a nematode-resistant rootstock. It is also phylloxera-resistant. This rootstock is not recommended for dry-land conditions, but is well suited to humid, compact, calcareous clay soils and to climates in the lower temperature range.
- **775 P, 779 P:** These are very valuable rootstocks, and their performance should be used on extremely difficult soil (such as compacted clays, water-logged, or shallow soils) and in arid conditions. 775 is too vigorous for fertile

soils and is accompanied by poor fruit set, but it is phylloxera-resistant. Can be planted at close spacing.

- **Freedom:** This rootstock is a good choice if you have sandy soil, and it is very vigorous in loam soils. It is phylloxera- and nematode-resistant.
- **SQR:** Good cold-climate rootstock; early ripening. Resistant to phylloxera.
- **SO4:** Vigorous rootstock with good phylloxera resistance and some nematode resistance. Good for moist conditions and clay soils; not drought-tolerant.
- **3309 C:** Good vigor and phylloxera resistance. Early bud break and ripening. Deep, well-branched root system. Prefers cool sites and close spacing.
- **101-14:** Average vigor with high phylloxera resistance. Tolerates wet soils but not drought; needs irrigation in hot climates. Early bud break. Good for cold climates and soils with low fertility.
- **Riparia gloire:** Favored for vigorous sites; early ripening and short growing cycle. Shallow roots best suited to deep, moist, fertile soils. Needs irrigation.

THE TRELLIS SYSTEM

The trellis system is the training and support system for your vines and also for the irrigation system. Historically, vines were planted without any support system, with wide spaces between vines, and without a supplementary water source. These were known as dry-farmed vineyards. Nowadays, the trellis system is an essential key to a successful vineyard. It supports the vines as they grow, allows you to manage the canopy to maximize sunlight to the developing fruit, and allows you to easily mow and maintain the space between the rows.

In most areas of the world, the vertical shoot positioning (VSP) trellis system for training your vines has proven to be the easiest to construct and understand, the

The Vertical Shoot Positioning (VSP) System

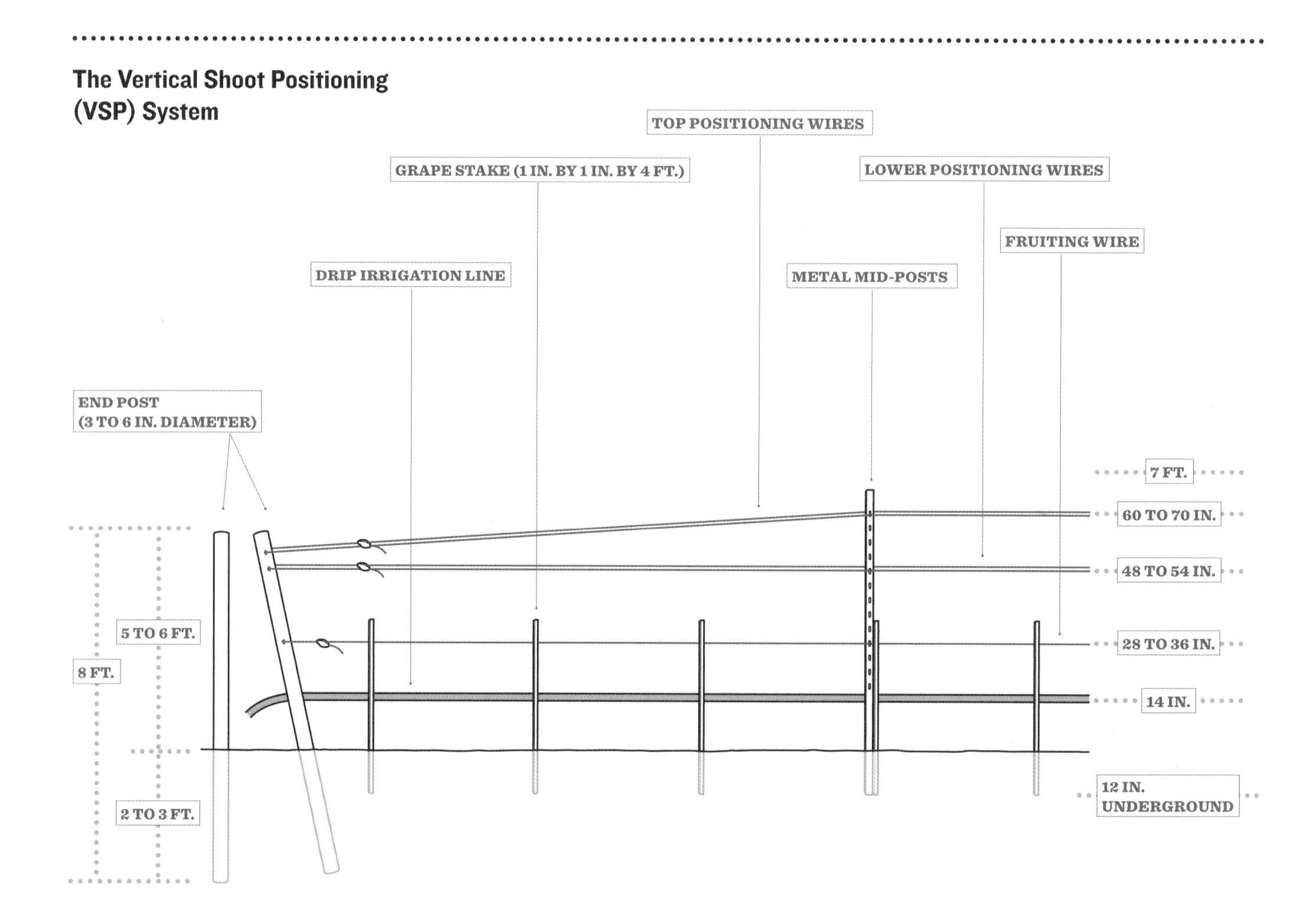

least expensive for wine grapes, and compact enough for a small grower. It is widely used in commercial vineyards, but can be adapted to the backyard, and small or large vineyards as well. You can plant the rows as close as is practical, and the vines will get enough sun to allow the photosynthetic process to produce fruitful grapes.

Each row of a VSP trellis consists of four rows of wire tightly strung between supporting posts. Heavy end posts provide the main support for the wires; lighter mid-posts serve as wire guides. The end posts are typically wood, but unless they are chemically treated, they must be a naturally rot-resistant wood such as cedar, juniper, or redwood. An alternative—which may be a requirement for organic certification—is to use metal end posts. There are a number of different types of metal posts. The most important consideration is whether they have sufficient tensile strength for the length of your rows.

The wire closest to the ground supports a drip irrigation tube at about 14 inches. Above this is the fruiting wire, a single wire where the grapes will develop. Above the fruiting wire are two double sets of positioning, or training, wires at approximately 4 and 5 feet. Eventually, when the vines are mature, the trellis must carry the weight of all the vines, irrigation lines, and (hopefully) many bunches of grapes. That is why the trellis system must be sturdy, well-designed, and properly installed.

BELOW: There are some slight variations on the VSP trellis system, but all have the same components: four rows of wires strung between end posts and supported by mid-posts. Some vineyards use metal end posts. BOTTOM LEFT: in my own vineyard I use wooden end posts and metal mid-posts. Depending on the length of the row, end posts may need additional support or bracing. The lowest wire supports the drip irrigation line.

You will plant your grapevines in the ground directly below the wires, each vine supported by a grape stake. As the vine grows, you will tie it to this stake, and it will develop into the main trunk. When the vine reaches the fruiting wire, you will train it in a horizontal direction on either side of the trunk by securing it to the wire to form two cordons, or arms.

As the cordon's shoots develop, you direct them vertically up between the shoot positioning wires (once the shoots start to grow, they are called canes). You will maintain the canopy of foliage between the positioning wires. As the plant matures and forms grapes (third or fourth year and beyond), the grapes form near the base of the canes in the fruiting zone above and below the fruiting wire. Each season you will adjust the height of the positioning wires to suit the growing canopy. Wire tightening may also be part of the yearly maintenance program, as is checking on all the elements of the trellis to ensure it is in good shape.

There are many places you can get the components for your trellis system, from building supply stores in areas where commercial vineyards exist, from farm supply stores, or from specialty mail-order suppliers.

Vine Growth on the VSP Trellis

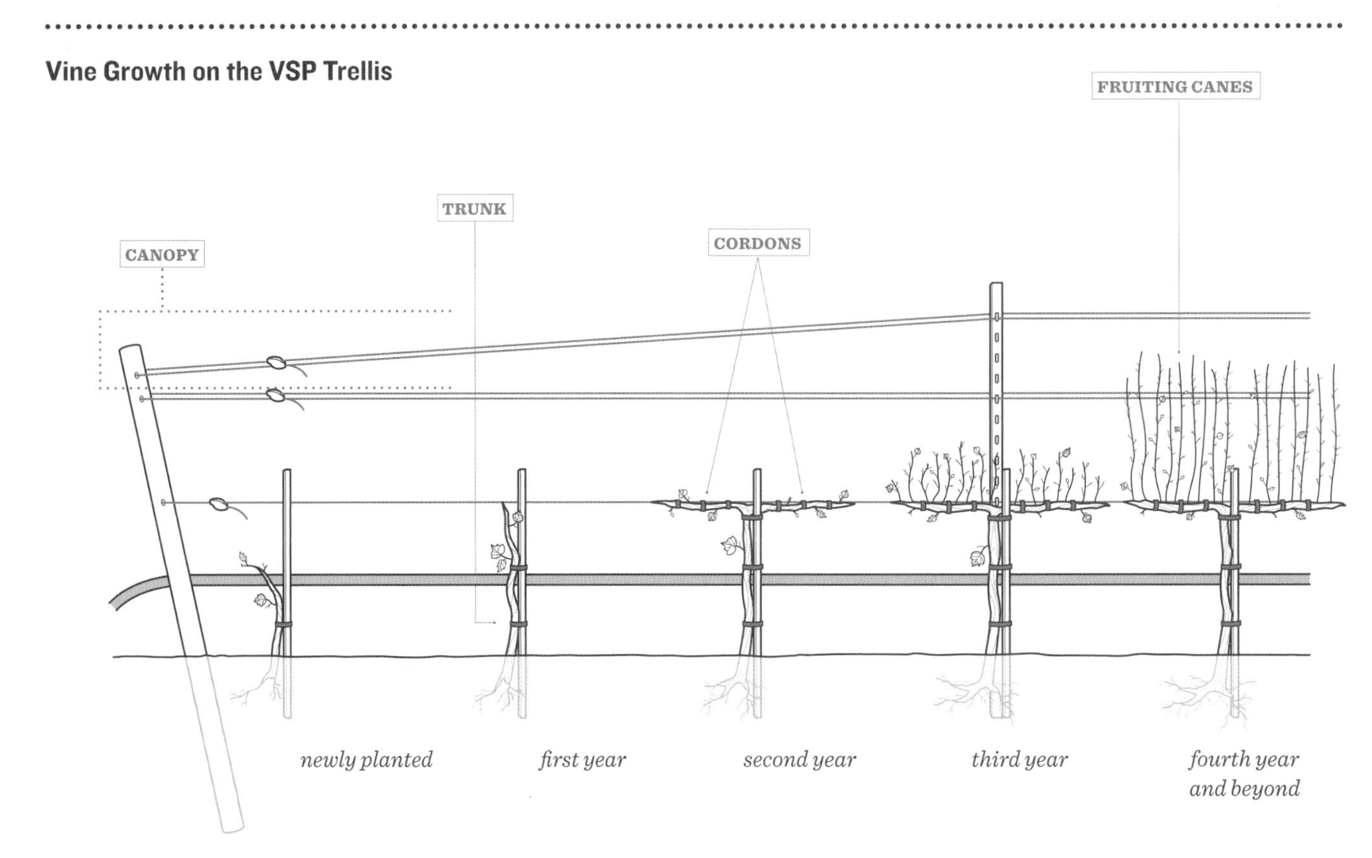

VINEYARD **IRRIGATION**

In their wild form, grapevines are fairly drought-tolerant, and in some regions, vines require only small amounts of irrigation as a supplement to rainfall and groundwater. Factors such as your climate, the vine spacing, and your average rainfall all influence the amount of irrigation water you will need, but most home vineyards in most places will require irrigation. However, one of the big problems I have seen with home vineyards is overwatering. You cannot water your grapes like you do your lawn or garden.

Although most municipal water sources are generally good, you might want to ask the local water service for the mineral content of the water supply. If you have a well, and have used the water to successfully grow other food or ornamental plants, you are probably not going to need a test for mineral content.

Water in the Vineyard

Water is used up in the vineyard through two different processes collectively known as evapotranspiration:

- Evaporation of water from the soil surface due to temperature and humidity levels.
- Evaporation of water especially from the underside of the leaves in the process called transpiration.

Both of these are regulated by your local climate and weather. Obviously, in hot areas there is more water loss from both evaporation and transpiration than in cooler climates. Also keep in mind that different soil types have different water needs, because different soil types hold water at different rates. Clay soils hold water for longer periods of time than loam, and sandy soil holds water for shorter periods of time than loam.

The ideal irrigation system consists of underground PVC pipes that lead to drip irrigation tubes on the trellis system. (Overhead watering is never recommended for a vineyard, as it can promote fungal diseases.) It should have a timer that allows you to set a water schedule that fits with your climate and soil conditions. The best time to install the irrigation system is when you are building the trellis before you plant, because young, newly grafted vines need sufficient water to develop healthy, vigorous root systems in the first year.

Managing vineyard irrigation can seem kind of tricky, but it's really not that bad. The general ideas are pretty simple: give the vines a good amount of water in the early years, and then not much later on. The goal is to maintain the moisture content

Connecting the Drip Line

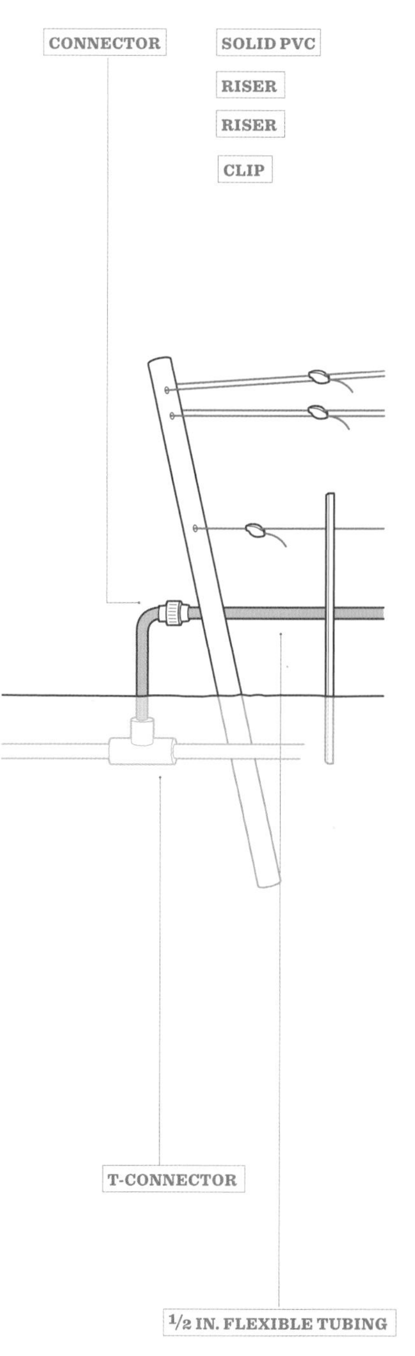

The drip irrigation tubing is secured to the bottom wire of the trellis system. Each vine has one dripper or emitter located above the root zone. Careful water management ensures that your vines do not get too much nor too little water at critical times such as fruit set.

of the root zone at a level throughout the sensitive stages of growth. Too little water for the vines early in the season can result in decreased vegetative growth. Too little water post-veraison can result in exposed fruit, shriveling, and lower yields. On the other hand, too much water damages roots by depriving them of oxygen, encourages excessive vegetative growth, and can increase the incidence of bunch rot and other fungus diseases.

The method of water management that I recommend is the water budget method, which is based on replacing the amount of water lost from evapotranspiration. However, you never allow the entire water reservoir in the soil to be lost. Because adverse plant stress in vines occurs after 50 to 60 percent of the soil water reservoir is lost, you replace the water budget in the root zone when water reaches the 50 percent moisture level.

Unless you have a lot of experience with managing an irrigation system, you may want to get a soil tensiometer to help you evaluate how much moisture is in the soil. If the soil is too dry, tension increases. If soil is too wet, its tension is low. Models are relatively inexpensive and have probes that go from a few inches into the ground to more than 3 feet in depth.

Tensiometer

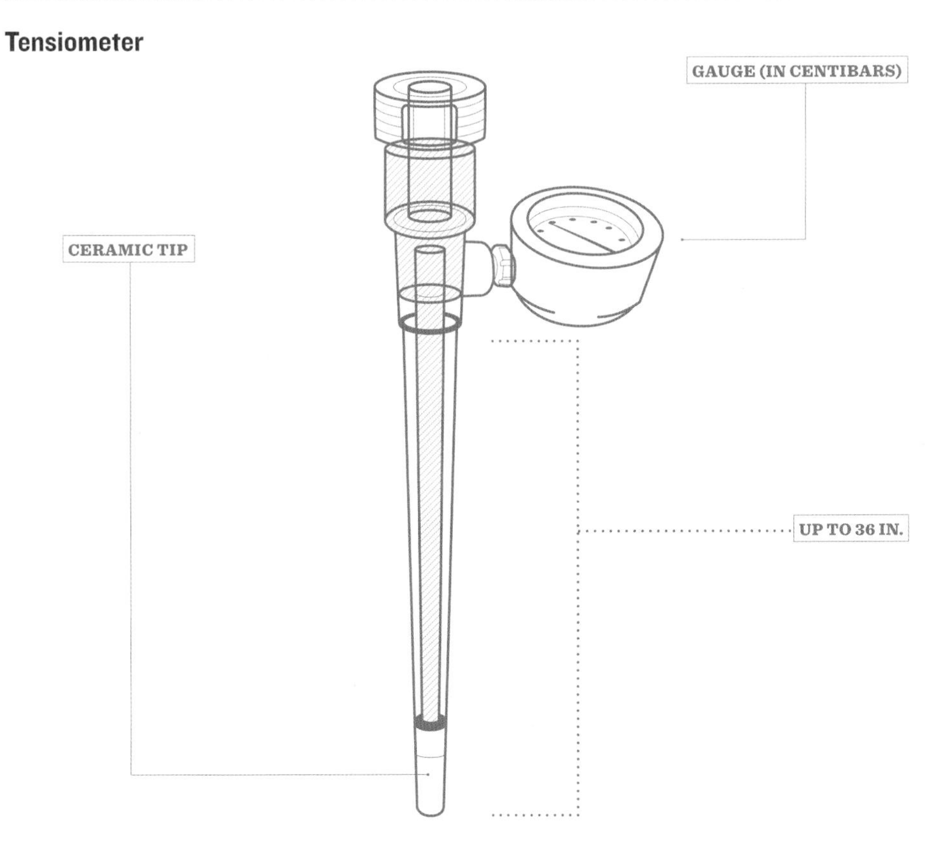

Careful management of irrigation water around the time of fruit ripening results in grapes with concentrated flavors.

Installing YOUR VINEYARD

INSTALLATION TIMELINE

CLEARING THE LAND

PREPARING THE SOIL

LAYING OUT THE VINEYARD

BUILDING THE TRELLIS

POST-PLANTING CARE

INSTALLATION TIMELINE

It is possible to plant any time during the growing season, but I recommend planting in early spring. This gives the plants plenty of time to establish their root systems over the summer. Because you are planting an organic vineyard, it's best to begin preparations a full year beforehand (assuming you don't already have a well-prepared, weed-free site). This will ensure that your soil is in optimum condition and that all perennial weeds have been killed. The work done in advance will pay off in the long term.

→ **Winter/spring one year before planting**
- Remove trees and shrubs, if needed.
- Remove lawn grass, if needed.
- Start to control perennial weeds.
- Plan vineyard layout.
- Check with the nursery and, if necessary, order vines now for planting the following spring.

→ **Spring/summer before planting**
- Do soil tests.
- When soil is dry, till or disc soil, removing rocks and stumps.
- Address any drainage issues.
- If necessary, solarize to remove annual weeds (four to six weeks).
- Continue tilling or discing to remove weeds.

→ **Fall before planting**
- Disc or till in amendments if indicated by the soil test.
- If vines were not ordered in spring, order them now.
- Build deer fence, if needed.
- Start building irrigation system: main lines, valves, and risers.
- Seed cover crop to improve soil and suppress weeds.

→ **Early spring**
- Mow or disc in cover crop.
- If soil is hard, soak it well at the post positions the day before installing the trellis.
- Install trellis: end posts, wires, mid-posts.
- Install drip irrigation lines and emitters; connect to risers and ensure irrigation system is working.
- Install grape stakes.
- After all danger of frost has passed, plant grapes and protect with grow tubes.

Young grapevines start to emerge from their grow tubes within a few weeks of planting in the new vineyard. This vineyard uses wood, not metal, mid-posts.

CLEARING THE LAND

In an organic vineyard, weeds are often the worst enemy. Careful preparation of the soil before planting will pay off in the long run, but remember that weed control will be an ongoing task. Most important is to remove perennial weeds—those that will grow back from the roots if they are not completely destroyed. Persistent culprits include Himalayan blackberry, Bermuda grass, and bindweed. No matter what your method of control for perennial weeds, it will probably take time and patience.

Studies by the U.S. Department of Agriculture indicate that a mixture of 25 percent vinegar and water will kill many weeds. My experience is that this mixture will work, but only for young weeds less than 6 inches in height. There are also a number of commercial organic products for weed control, but again, they seem to work best only on smaller weeds such as spurge, sowthistle, shepherd's purse, London rocket, fleabane, and some grasses.

If you are mainly concerned with annual weeds, you can cultivate the soil by discing or tilling at intervals over several weeks. Irrigate the soil between cultivating to encourage weeds to sprout, then cultivate again. Gradually you will reduce the amount of weed seeds in the soil. Make sure to do this before the weeds set seed, or you are just helping them to grow.

A rototiller is invaluable when preparing a small vineyard. You can use it to help clear weeds, break up compacted earth, and incorporate any needed amendments or compost into the soil before planting.

If you have a large area of grass or weeds, you can kill it by solarization, a method that requires some patience and plenty of hot sun. Place clear construction-grade plastic sheets (typically 6-ml) on the planting area and secure them with heavy rocks or pegs. Leave for up to six weeks. The temperatures will rise under the plastic and kill not only weeds, but many soil pathogens as well.

Another way to clear an area of weeds is to rent or borrow some sheep or goats to do the job for you. Again, this is a process that will take time, and you will still need to cut down larger weeds that are too tough for the animals to devour. This works well to completely clear an area of unwanted plants.

I let sheep into my vineyard every winter, after leaf fall. They keep down the weeds and at the same time they produce fertilizer. Goats are a little too aggressive in my experience to let loose in an existing vineyard. I have seen them eat dormant vines. They also have a tendency to climb up on things to get at trees near the vineyard, whereas sheep never leave the ground. However, goats can be useful if you are trying to clear the land of weeds before planting.

Sheep and goats make good weed-control devices. Although a goat can clear weeds pre-planting, I prefer to let only sheep into the vineyard once it is planted. They will take care of the weeds but leave the dormant vines alone.

FENCING ISSUES

If deer roam near your property, you can be sure that they will be pests in the vineyard. The single best solution for deer is to build a fence to keep them out, and the time to do this is before you plant. Although wire fencing is the best and most durable, a cheaper option is a polypropylene mesh fence. Whatever fencing material you use, a deer fence should be at least 7 feet high.

In some neighborhoods, there is a limit on how high you can build a fence, but you may be able to top a wooden or chain link fence with a row or two of wires.

PREPARING THE SOIL

Good soil preparation is essential for the vineyard. If you have done a soil test and it indicated that you have a nutrient deficiency or pH that is too high or low, you will need to amend the soil before planting, according to the recommendations from the soil test. These recommendations will tell you how much amendment to dig into the soil by area. You might also want to dig in some rotted manure or compost the fall before planting, even if your soil has no deficiencies.

In general, the best time to apply amendments is in the fall or in the early spring when the ground is easiest to turn over. Spread the amendment, manure, or compost, and then rototill or disc over the ground to no more than 4 to 6 inches deep. Cover crops like fava beans and other legumes should then be planted so that winter and spring rains can establish them.

LAYING OUT THE VINEYARD

There are two main measurements for the vineyard. The first is the amount of space between the vines in each row: typical spacing between plants is 4, 5, or 6 feet. The second is the spacing between rows—usually 6, 7, or 8 feet. In fertile soils, rows and plants can be spaced closer together; your vines will still be productive, but your vineyard will cost less to install. If you can go to 5-foot plant spacing and 7-foot rows, do so. I believe a vineyard with plants spaced at 5 or 6 feet and 8-foot rows is best if you have fertile soils and a vigorous rootstock such as 110R.

This means that there are different dimensions for any vineyard. The chart shows some alternate spacing for an acre (208 by 208 feet), 1/4 acre (approximately 100 by 100 feet), and 1/16 acre (approximately 50 by 50 feet), and how many vines you will need to order if using that spacing.

Some varieties are more productive than others, just as some rootstocks are more productive than others. So if you are planting a variety known to be highly productive, you might space the vines further apart (because you'll get more fruit per plant) or put it on a rootstock that would temper its vigor. Each situation is going to be somewhat different depending on soil and climate.

Again, rows can run east-west or north-south. If your land is sloping, vines lower down the slope will be more subject to frost damage. Also, if your vineyard is on a hill with a slope up to 30 percent and you want to use wheeled vehicles for maintenance, you should make the rows run up and down the hill, not across it.

To lay out the vineyard, you can use stakes and string, spray paint, or lime to mark the position of the rows. Begin with a base row, then mark each row off this base row. You can use the 3-4-5 rule to make sure that the ends of each row are perpendicular to the base row. Measure this carefully so that your rows are correctly spaced and straight. For a large vineyard, you can use a transit to determine exact measurements and align the rows.

The installation shown in our illustration is based on constructing a 1/16-acre vineyard (approximately 50 by 50 feet), with vines spaced 5 feet apart and rows spaced 7 feet apart.

VINE AND ROW SPACING

VINEYARD SIZE	VINE SPACING	NUMBER OF VINES			
ROW SPACING		6 FT.	7 FT.	8 FT.	9 FT.
1 acre	4 ft.	1815	1556	1361	1210
1/4 acre	4 ft.	454	389	340	302
1/16 acre	4 ft.	113	97	85	76
1 acre	5 ft.	1452	1245	1089	968
1/4 acre	5 ft.	363	311	272	242
1/16 acre	5 ft.	90	72	68	60
1 acre	6 ft.	1210	1037	908	807
1/4 acre	6 ft.	302	259	227	202
1/16 acre	6 ft.	76	65	57	50
1 acre	7 ft.	1037	889	778	691
1/4 acre	7 ft.	259	222	194	173
1/16 acre	7 ft.	65	56	49	43
1 acre	8 ft.	908	778	681	605
1/4 acre	8 ft.	227	194	170	151
1/16 acre	8 ft.	57	49	42	39

A Sample Site Plan for a Small Vineyard

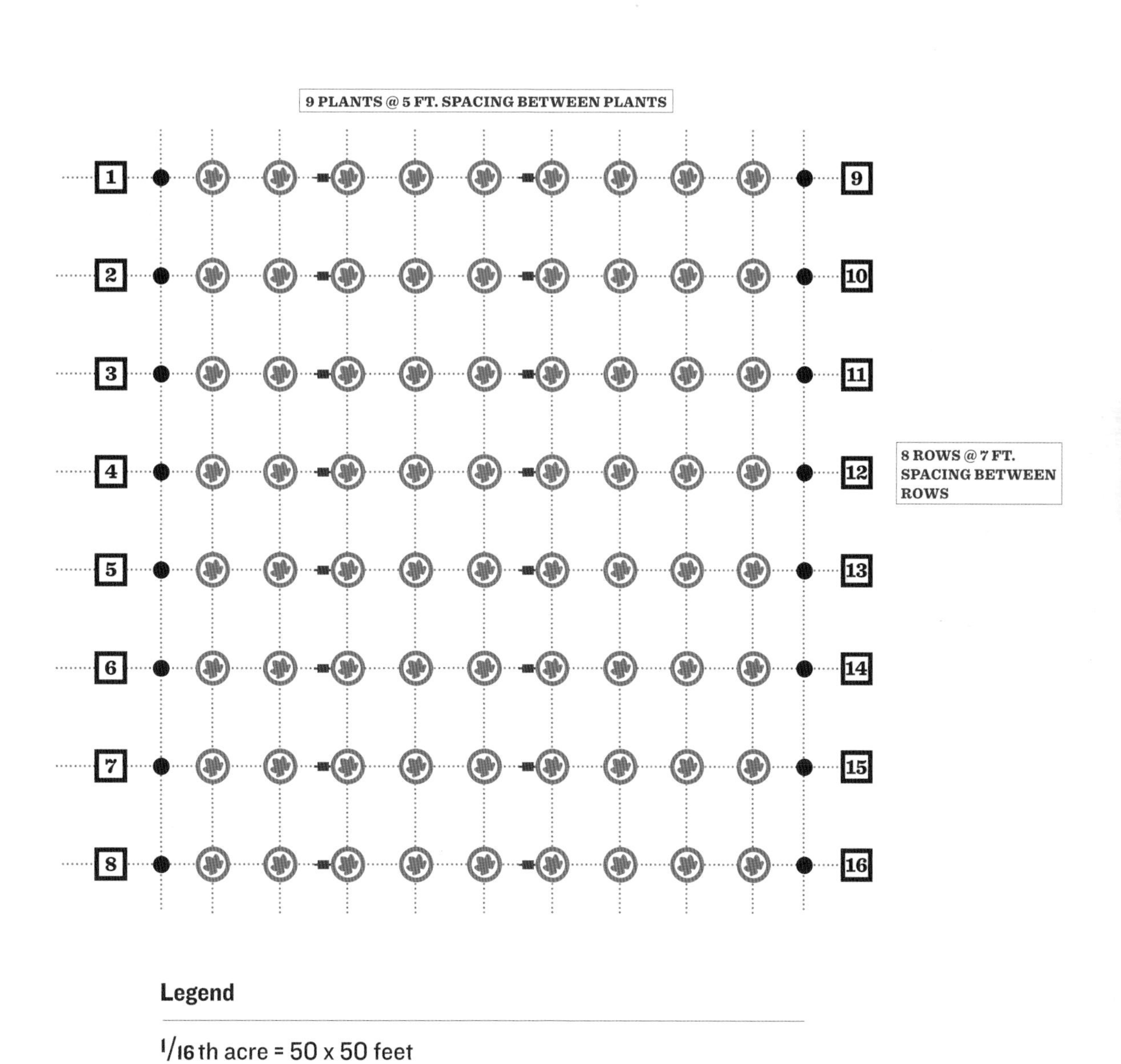

Legend

1/16th acre = 50 x 50 feet

● → 16 end posts | → 72 plants | → 16 mid-posts

TOOLS AND EQUIPMENT

Before you start on vineyard construction, you'll need to make sure you have the correct tools. Chances are you have many of the common home and garden tools shown in the illustrations. However, there are some special tools that you may need to purchase or rent.

- **Post-hole digger:** End posts need to be at least 2 feet deep, and if the rows are longer than 200 feet, they should go down 3 feet in depth. You can also rent a power soil auger or hire a someone with a Bobcat auger to do the job of digging holes for your posts.
- **Metal post driver:** A sturdy metal tube with handles that allows you to pound the posts into the ground. These work well and are safer than using a sledgehammer.
- **Fencing tool:** You can find this tool at hardware stores where fencing is sold. It is a combination of a small hammer, wire cutters, and pliers built into one tool. You will need it to pull the wire, twist it, and cut it.
- **Wire cutters:** Strong wire cutters that can cut up to 10 gauge wire will come in very handy. You'll find them at hardware stores or from a vineyard supplier.
- **Grippler tool, Gripples:** This specialized tool secures and tensions the wires in the trellis system. Stiff high-tensile steel wire is used for vineyard trellises, and this tool is used to fasten and adjust the proprietary Gripple joiners and tensioners. The Gripples can be adjusted and re-tensioned year after year. If you buy a large box of Gripples, the tool is often included.

Construction Tools and Equipment

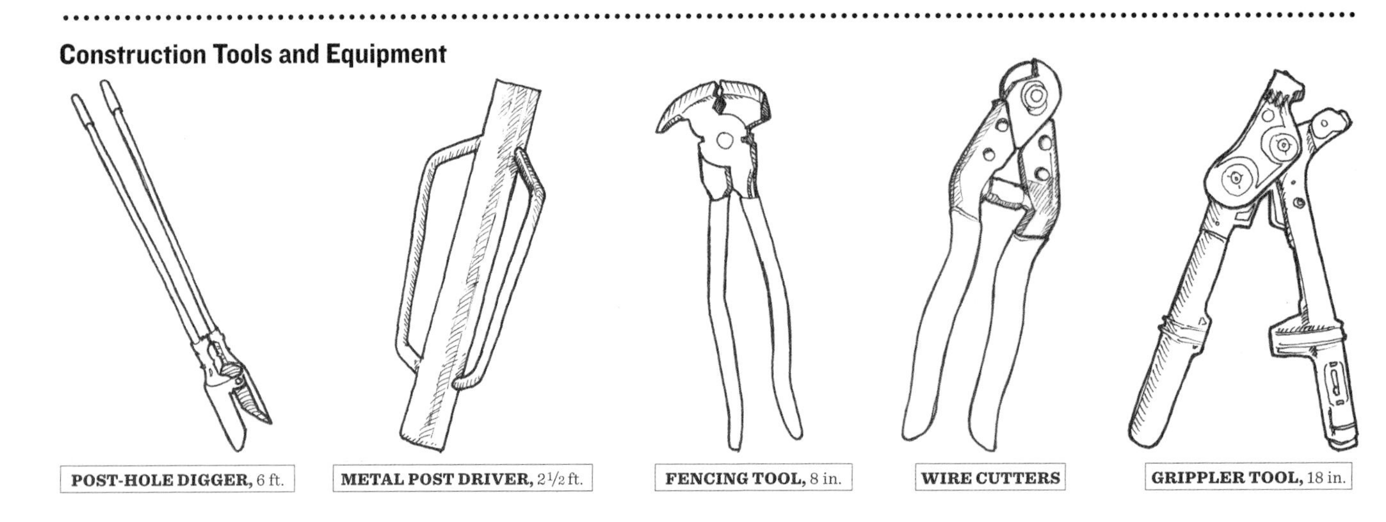

POST-HOLE DIGGER, 6 ft. | METAL POST DRIVER, 2 1/2 ft. | FENCING TOOL, 8 in. | WIRE CUTTERS | GRIPPLER TOOL, 18 in.

Supplies

For the trellis

16 wood posts, 8 feet long and 3–4 inches in diameter
16 metal mid-posts
72 wood grape stakes (4 feet long by 1 inch square)
3000 feet of 12 or 14 gauge, high-tensile wire
50 wire vices or
50 Gripples
70 wire clips

The diameter of wooden end posts should be determined by the length of the row—that is, a row of 20 to 40 feet would only require 3- to 4-inch diameter end posts, while a row of 50 to 100 feet would need 5-inch end posts and a row of greater than 100 feet would require 6-inch end posts.

For the wire, the larger the gauge, the thinner the wire. Use 14 gauge wire for shorter rows, say, under 100 feet, and 12 gauge for longer rows. I would use 10 gauge for the fruiting wire in very long rows of over 500 feet.

Construction Supplies

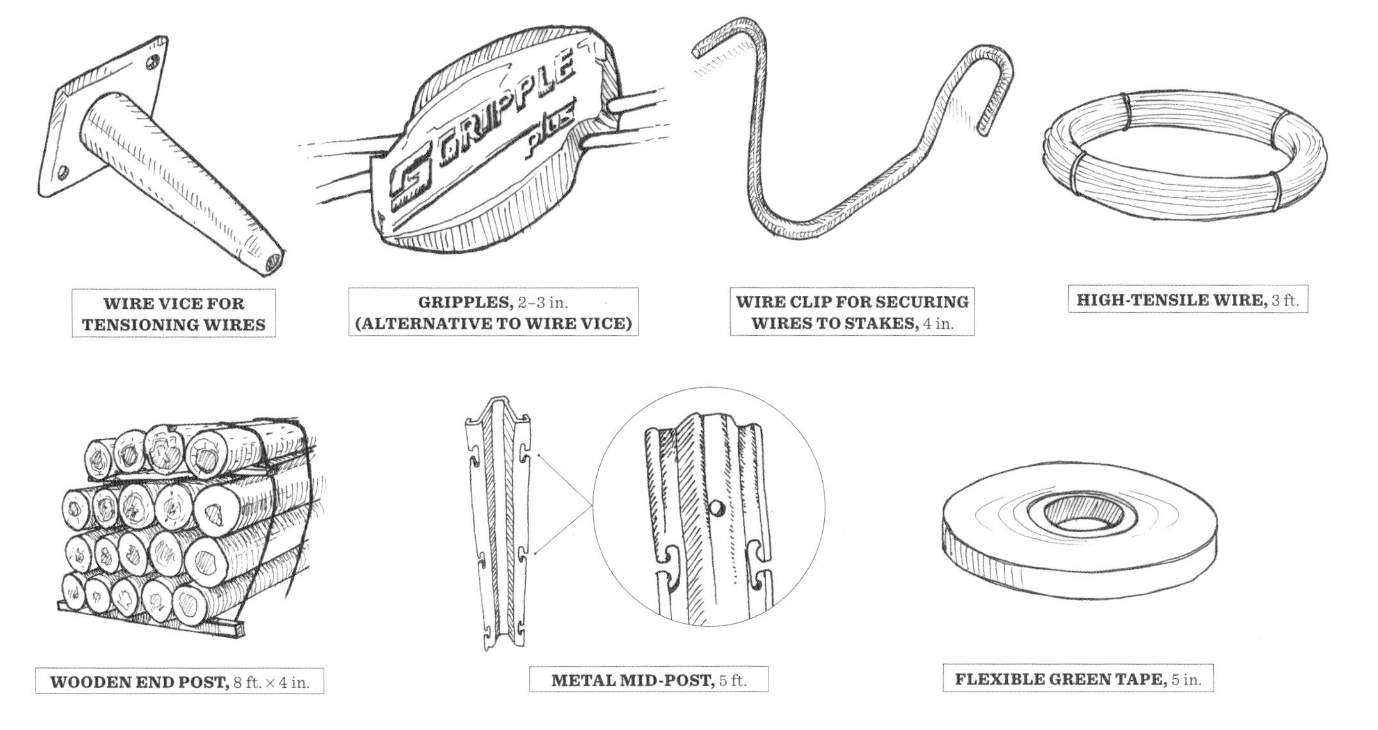

WIRE VICE FOR TENSIONING WIRES

GRIPPLES, 2–3 in. **(ALTERNATIVE TO WIRE VICE)**

WIRE CLIP FOR SECURING WIRES TO STAKES, 4 in.

HIGH-TENSILE WIRE, 3 ft.

WOODEN END POST, 8 ft. × 4 in.

METAL MID-POST, 5 ft.

FLEXIBLE GREEN TAPE, 5 in.

A WORD ABOUT GROW TUBES

Plastic grow tubes create a mini-greenhouse around the young vine that promotes vine growth and encourages the development of a single strong shoot that will become the main trunk. They also protect the young vines from sunburn and wind damage, and from nibbling animals such as rabbits. You can purchase grow tubes from a vineyard supplier, but as they are reusable, you can often purchase them second-hand from a commercial winery. Other grow tubes are made so that you form them into a triangular or cylindrical shape. Some growers still use paper milk cartons with the top and bottom cut off. The tubes are usually removed at the end of the first year's growth.

Grow tubes help protect young vines from damage and encourage faster shoot development.

For the irrigation system

500 feet of ½ in. flexible tubing
72 ½-gallon-per-hour drippers
50 hose clamps or curly ties
Fittings: L-connectors, T-connectors, end plugs or clamps, plus a timer-to-hose faucet adapter and goof plugs
A timer and valve

Other

Gravel for filling posts holes
72 grow tubes
Flexible green tape

Irrigation Supplies and Tools

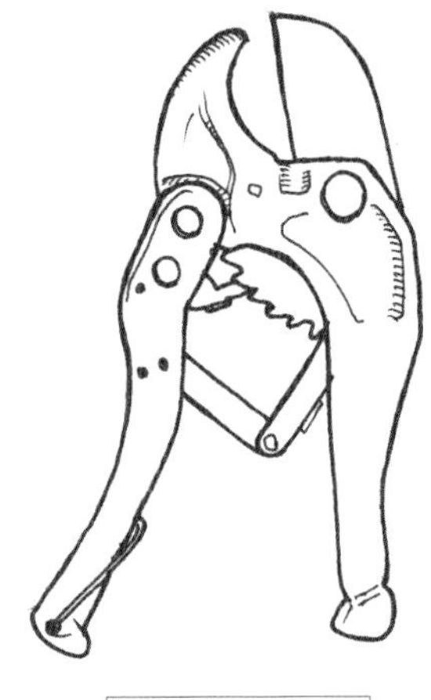

PVC PIPE CUTTER

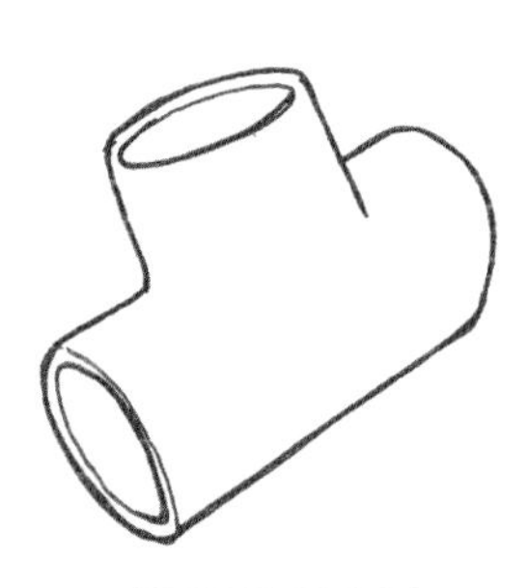

T-CONNECTOR

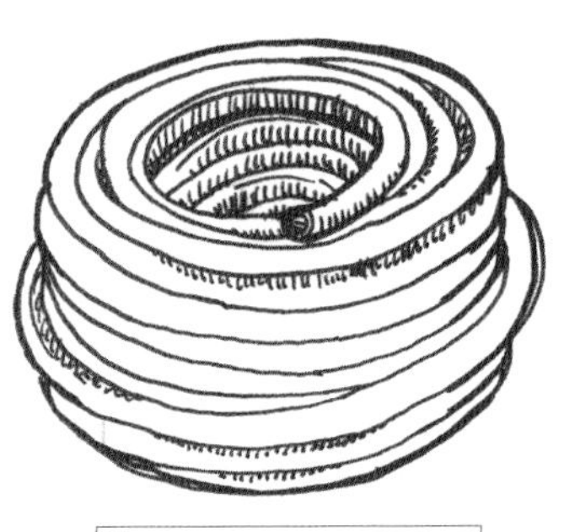

½ IN. FLEXIBLE TUBING

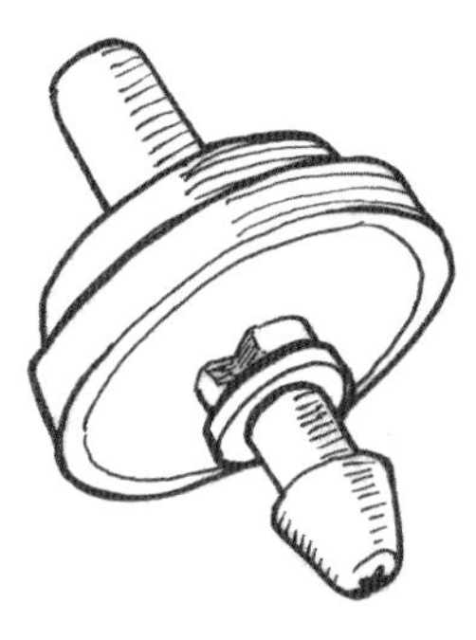

DRIPPERS

PVC GLUE

BUILDING **THE TRELLIS**

Begin by laying out all of the materials in position. Even if you have marked out the vineyard beforehand, measure and mark the location of the end posts again to be sure all your estimates are correct. See post numbers on page 71.

Installing the End Posts

1. Start with post #1.
2. Measure 7 ft. from post #1 to post #2, then 7 ft. to post #3, and then to posts #4, #5, #6, #7, and #8 in order.
3. Measure 49 ft. from post #1 to post #9.
4. Next, measure 7 ft. from post #9 for posts #10, #11, #12, #13, #14, #15, and #16, in order.
5. Dig all 16 post holes, 2 to 3 ft. in depth.
6. Place each end post in its hole. Hold each post in position so that it leans away from the center of the vineyard about 8 to 12 in. Fill in around the base of the post with gravel and soil.
7. Tamp the gravel and soil, so that the posts are secure.

Digging Post Holes

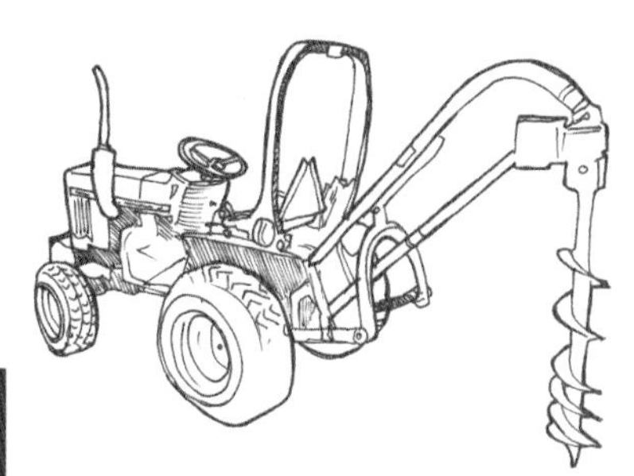

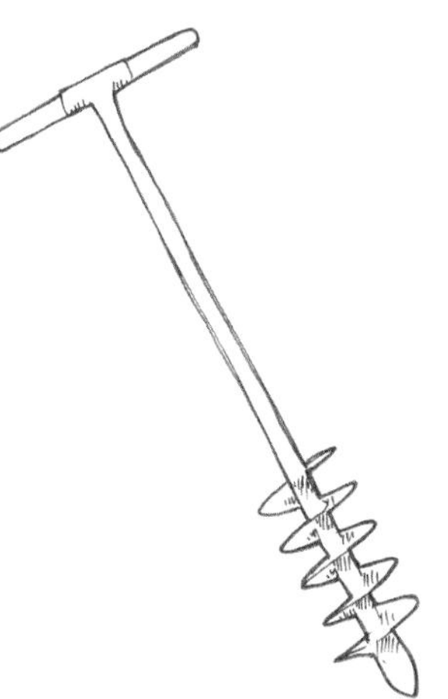

Your post holes should be evenly spaced and set in the ground to a depth of 2 or 3 ft. Renting a soil auger can make digging the holes for them a much easier job. Soil augers may be hand operated or fitted as an attachment to a Bobcat or small tractor.

Drilling Wire Holes

Note: The exact height of the fruiting wires should be determined by the height of the person who will be pruning and tending the vines.

1. Measure 14 in. above the ground and drill a 1/4 in. hole through the center of posts #1 to #8 facing into the center of the vineyard. This will be for the bottom (irrigation) wire.

2. Measure 28 to 36 in. above the ground and drill another 1/4 in. hole through the center of end posts #1 to #8. This will be for the fruiting wire.

3. Measure down 2 in. from the top of each end post #1 to #8 and drill two 1/4 in. holes side by side, for the top shoot position wires.

4. Measure halfway between the fruiting wire and the top wires on each end post #1 to #8 and drill two more 1/4 in. holes for the lower shoot position wires.

5. If using wire vices (see step 3b in next section), drill the same set of holes in posts #9 to #16.

Installing the Wires and Irrigation Tubing

1. Cut the bottom wire for row #1 at least 60 ft. in length and place the wire in the bottom hole on post #1. Wrap one end of the wire around the post, thread it back through the wire in the hole, and twist it tightly with the pliers. This end of the wire will always be secured so you do not need to adjust it. You will adjust the tension at the other end of the row with a Grippler tool or wire vice.

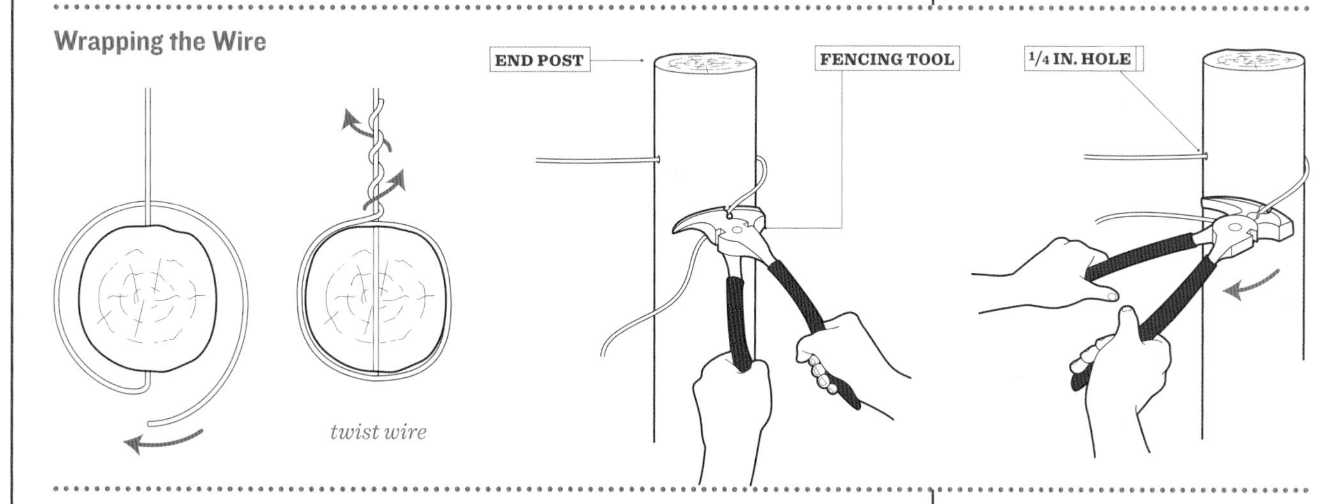

2. Repeat step 1 for the next seven bottom wires, then for the fruiting wires and then the two sets of positioning wires. Now all your wires are ready to be secured to the opposite posts.

continued on next page

→→
continued from previous page

Installing Wire Gripples

Thread the wire through one side of the Gripple, then wrap it around the end post and back to the Gripple. Thread the wire through the Gripple in the other direction. Insert the Gripple into the tool and open the handles.

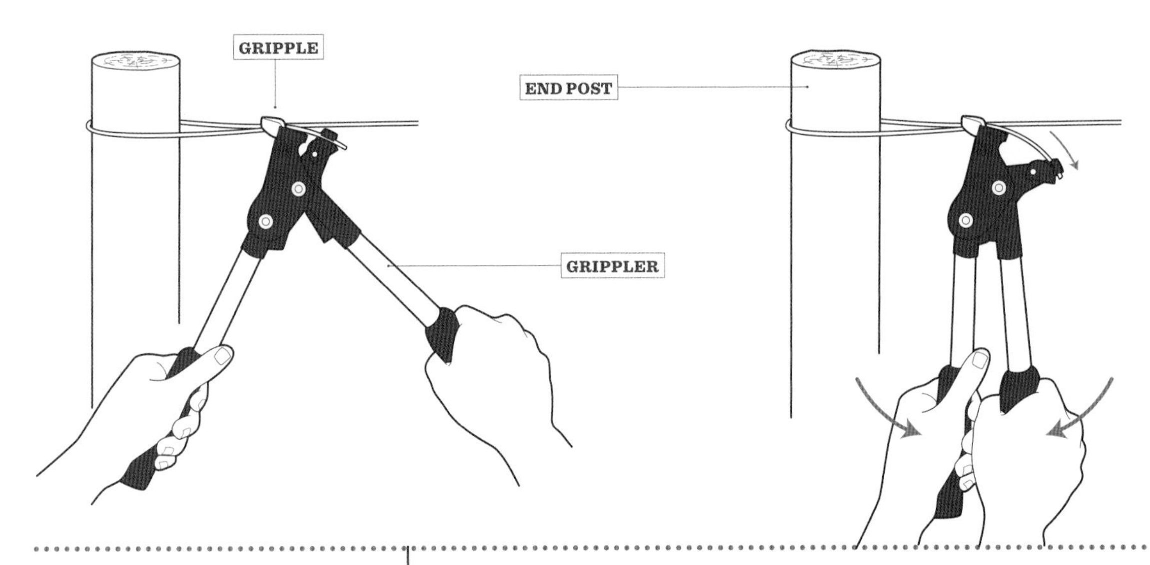

3a. If using Gripples, thread the end of wire through one hole in the Gripple, wrap the wire around the opposite end post at the correct height, and then thread it back through the other hole in the Gripple. Use the Grippler tool to tighten the wire.

3b. If using wire vices, thread the end of each wire through the post, then into the vice, making sure that the tapered end of the vice goes into the pre-drilled holes. Tension will hold the wire in place. Pull the wire tight through the wire vice with pliers or the fencing tool.

Installing Wire Vice

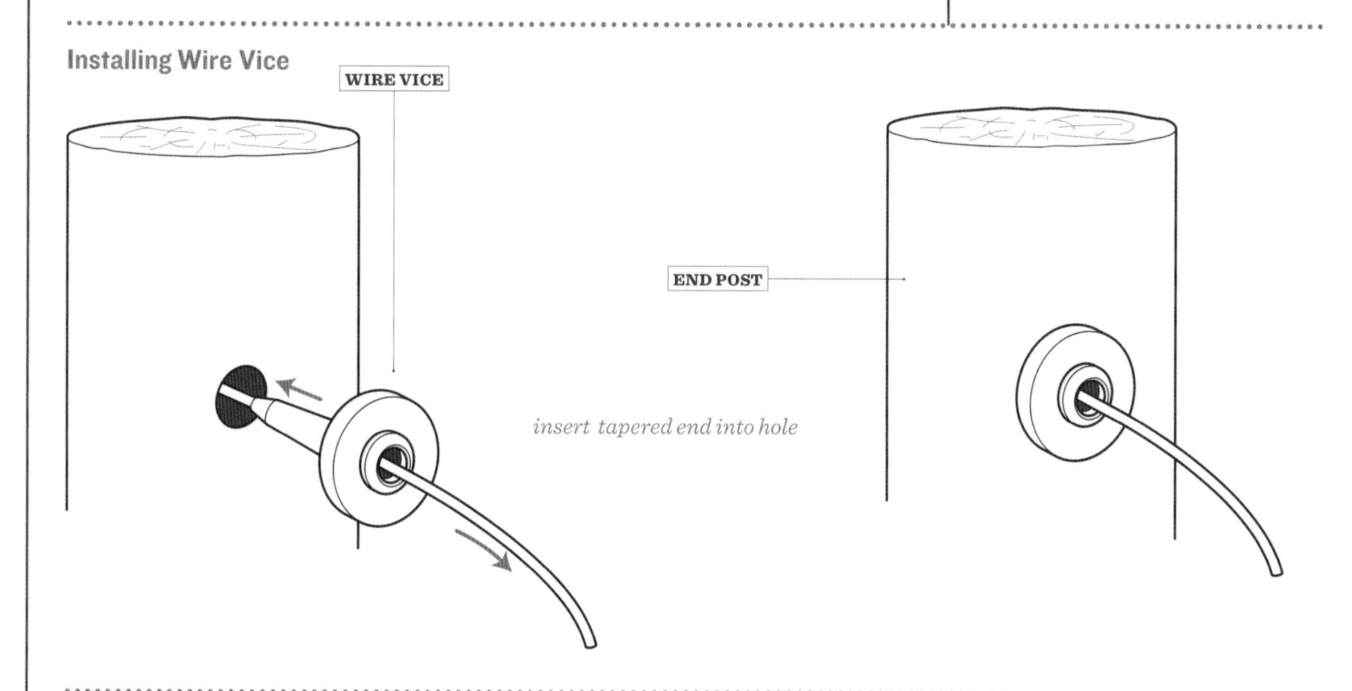

4. Cut one length of irrigation tubing 60 feet long and fasten it to the bottom wire between post #1 and post #9 with the hose clamps or curly ties, placing the clamps approximately 36 to 60 in. apart. Fit the tubing to the underground PVC riser at end post #1. Then put a hose crimp (or end plug) on the other end of the tubing, at post #9.

5. Repeat hose cutting and clamping of the tubing as above for each row.

6. Measure 2 ft. from end post #1 and place your first dripper. Then continue at 5 ft. intervals (according to your vine spacing) so that each vine will have one 1/2 gallon-per-hour (gph) dripper positioned directly over it.

Installing Grape Stakes and Mid-Posts

1. Measure 2 ft. from end post #1 and place your first grape stake (at the same spot as the first dripper). With a hammer, drive it into the ground about 6 to 12 inches, and using the metal clips, secure the stake to the drip line wire and fruiting wire (use a screwdriver to bend the clips around the wires).

2. Measure 5 ft. from the first stake to the second stake position and repeat step 1. Do the same for each row, one at a time, to ensure that the stakes are lined up row to row.

3. Position a wire guide post every third grape stake. Use a sledgehammer or a metal post driver to pound each post about 12 in. into the ground. Use a level to ensure that the post is upright.

4. Attach the wires to each grape stake and mid-post. Use a screwdriver to fasten clips to the grape stakes to hold the wires.

Fastening the Wires

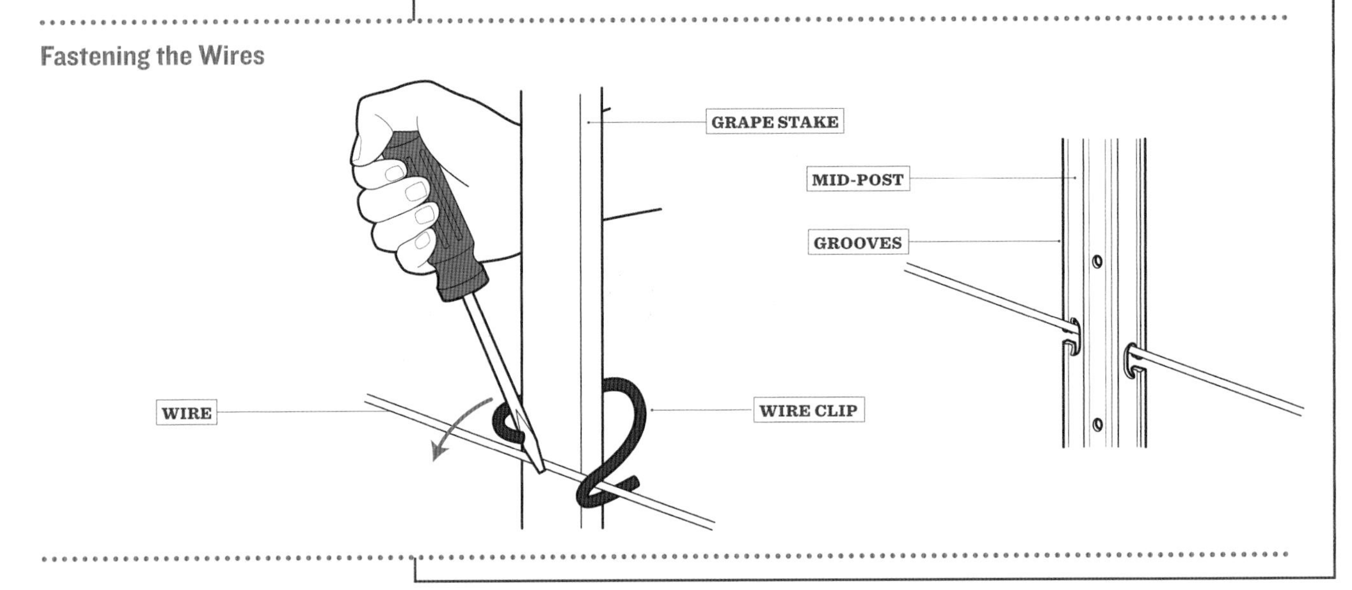

If you plant green-growing vines and they have more than one strong shoot, cut off the other shoots at planting time to encourage development of a single trunk.

Planting the Vines

1. Dig a hole at the base of each grape stake 8 to 10 in. deep and place the plant in the ground so that the graft union is about 3 to 4 in. above the surface of the ground.

Planting Depth

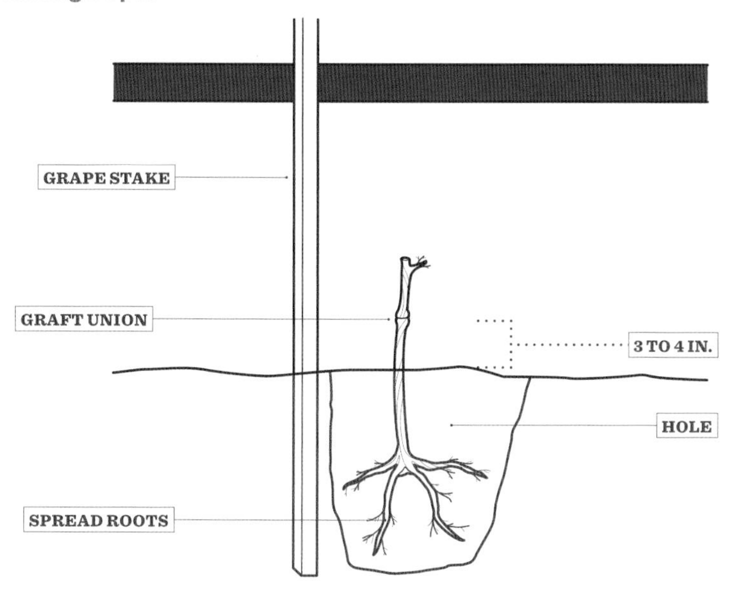

2. Fill the hole with the remaining dirt and gently tamp it with your foot or the shovel so that it is firmly packed.

3. If using grow tubes, fit a tube around the plant. Tie the tube to the grape stake with flexible green ties.

Step back and admire your work, and congratulate yourself and your partners on a job well done!

POST-PLANTING CARE

I like to water every day for the first six weeks, then back off to every other day for three weeks, then every third day for the rest of the first year. Use your judgment making sure that the weather is warm enough to dry out the ground around the drippers so the plants' roots are not always in wet soil. If you have a tensiometer, use it to ensure that the soil is not oversaturated. It is not necessary to water after the plants have become dormant.

Continue to check that the drippers are working and that all the wires are secure. Make sure that all mid-posts and stakes are upright. If you need to correct the position of a stake, now is the time to do it, before the vine roots have spread.

Keep an eye on weeds, especially those that are growing up around the young vines, including inside the grow tubes. Pull them by hand to avoid damaging the vine roots.

THE VINEYARD YEAR

MIDWINTER

VINE STATUS:
Dormant

IN THE VINEYARD:
- → In mild climates, keep down weeds.
- → In mild climates, maintain cover crop.
- → Apply dormant spray of lime-sulfur.

LATE WINTER

VINE STATUS:
Dormant

IN THE VINEYARD:
- → Prune vines to spurs with two growing buds.
- → Inspect the trellis wires and tighten as necessary to increase tension.
- → After pruning, apply lime-sulfur to dormant vines for anthracnose.
- → In warmer areas, monitor vines for early signs of pests like borers, mealybugs, and spider mites.

EARLY SPRING

VINE STATUS:

Bud burst

IN THE VINEYARD:

- Keep an eye out for swelling buds on the canes, which happens when the average daily temperature is 50°F (10°C).
- At bud burst, begin spraying for fungal diseases.
- Inspect the irrigation system and make sure the drip lines are clamped to the wire. Replace any damaged or missing drip emitters.
- Mow or till cover crop between rows.
- Watch for late frosts and protect the vines as needed.
- Remove winter-damaged wood, and any vines that are sickly or weak.
- In cold-winter climates, sow spring cover crop.
- Control emerging weeds around the base of the vines.
- Monitor vines and growing spurs for pests like mealybugs, spider mites, and thrips. Treat as needed. If sharpshooters or caterpillars occur in your area, put up sticky or pheromone traps.

MIDSPRING

VINE STATUS:

Flowering occurs when the average temperature is between 61°F and 68°F (from 30 to 80 days after bud burst).

IN THE VINEYARD:

- Monitor vines for pests such as thrips, spider mites, leafhoppers, mealybugs, and sharpshooters.
- A flush of root growth begins.
- Continue to spray for fungal diseases every week or two.
- Control weeds under the vines.
- Mow or till cover crop between the rows.
- Second year vines: Tie the shoots as they grow up the stake and out the fruiting wire.
- Apply a well-balanced fertilizer.
- Gophers become very active now; be sure to control them before they get out of hand.

LATE SPRING

VINE STATUS:

Fruit set comes 7 to 10 days after flowering. / Rapid shoot growth

IN THE VINEYARD:

- Monitor vines for Japanese beetles.
- Look for signs of eutypa or phomopsis.
- Continue to spray for fungal diseases every week or two.
- Control weeds under the vines.
- Mow weeds, cover crop, or ground cover between the rows.
- First year vines: Tie shoots as they grow out along the fruiting wire; these will become the permanent cordons
- Monitor the water needs of the vines, checking for signs of drought stress and checking soil moisture with a tensiometer.
- First and second year vines: Pinch off grapes.
- Third year vines: Pinch off grapes on vines that do not have bark on the trunk and cordons.

EARLY SUMMER

VINE STATUS:

Late fruit set / Rapid shoot growth

IN THE VINEYARD:

- Continue to spray for fungal diseases every week or two.
- Control weeds under the vines.
- Mow weeds, cover crop, or ground cover between the rows.
- First year vines: Continue to tie the growing shoots as they grow out along the fruiting wire.
- Third year and beyond: Start shoot training—pushing the shoots up between the double wires.
- Third year and beyond: Thin shoots and remove laterals.
- Remove suckers from the base of the vine.
- Continue to monitor the water needs of the vines, checking for signs of drought stress and checking soil moisture with a tensiometer.
- Third year and beyond: Trim shoots above the top of the second set of wires to 6 to 12 in.
- If the weather is dry, stop spraying for fungal diseases.
- Continue gopher control.
- Check fencing for holes or breaches that can let in deer.

MIDSUMMER

VINE STATUS:

Early veraison / Rapid shoot growth

IN THE VINEYARD:

- Control weeds under the vines.
- Mow weeds, cover crop, or ground cover between the rows.
- Continue to monitor for pests likespider mites, leafhoppers, sharpshooters, grass-hoppers, and Japanese beetles. Treat if needed.
- First year vines: Continue to tie the growing shoots that grow out along the fruiting wire.
- Third year and beyond: Continue shoot training—pushing the shoots up between the double positioning wires.
- Third year and beyond: Remove laterals.
- Third year and beyond: Thin the leaves around the fruiting zone to leave 70 to 80 percent of the fruit exposed.
- As veraison starts, turn off the irrigation.
- In clay soils, if the ground is cracking, till or disc to eliminate cracks.
- Keep an eye out for birds and make sure you have sufficient netting in stock.
- Continue to control gophers, and new pests like raccoons or foxes that may come into the vineyard attracted by the ripening fruit.

LATE SUMMER

VINE STATUS:

Shoots growth slows / Late veraison / Grapes turning to their final color

IN THE VINEYARD:

- Mow weeds, cover crop, or ground cover between the rows.
- Continue to monitor for pests like spider mites, leafhoppers, sharpshooters, and grasshoppers. Treat if needed.
- If rain starts after veraison, look for early signs of bunch rot.
- Third year and beyond: Remove laterals and suckers.
- Third year and beyond: Continue to trim shoots above the top positioning wires.
- Third year and beyond: Continue to thin leaves around the fruiting zone.
- Begin to test Brix with the refractometer.

FALL

VINE STATUS:

Shoot growth slows. / Grapes ripen. / Period of root growth begins.

IN THE VINEYARD:

- Continue to test Brix with the refractometer.
- Test grapes for pH.
- Be extra vigilant looking for animal pests like raccoons, foxes, and rabbits; they are attracted to ripe grapes.
- Put up netting for grapes.
- Harvest grapes.
- Give vines a deep watering after harvest, then stop watering.
- Put grape pomace into the compost pile unless grapes have mealybugs.

LATE FALL

VINE STATUS:

Leaves turn yellow, red, and eventually brown.

IN THE VINEYARD:

- If there are diseases or pests in the vineyard, pick up fallen leaves and dispose of them.
- Sow winter cover crops.

WINTER

VINE STATUS:

Root growth stops.

IN THE VINEYARD:

- First three years: Apply a balanced fertilizer.
- Four years and beyond: Decrease fertilizer based on vigor of plants.
- Clean up fallen bunches and clusters still on the vine.
- Harvest icewine grapes.

4

ALL ABOUT PRUNING

FIRST *YEAR*

SECOND YEAR

THIRD YEAR

Pruning AND Training

FOURTH *YEAR*

FIFTH YEAR AND BEYOND

MANAGING EXCESS VIGOR

The goal in the first few years is to train the young vines to grow a straight, strong trunk with two cordons, or arms, on either side.

ALL ABOUT **PRUNING**

Throughout the year, you will be tending to your vines in order to manage their growth. If you follow the steps of pruning and training, you'll maximize photosynthesis, encourage strong growth, and help your vines to produce the best possible fruit.

Your goals in pruning and training are to:

- maximize exposure of the leaves for photosynthesis
- control the vigor of the vines
- control the amount of fruit
- control the quality of the fruit
- encourage even ripening of the fruit
- allow air circulation to prevent fungal diseases
- make the fruit accessible for harvesting

Our goal in working with the vines is, ultimately, to develop two cordons per plant, with about 10 to 15 spurs on each cordon. Each spur should have two buds after pruning. In the spring, these spurs will be the basis for shoots that will be trained up into the positioning wires as canes—the grape-producing parts of the vine.

Remember that photosynthesis during the growing season makes carbohydrates that are stored in the plant to supply food for the next year's growth. So during the growing season, you want to promote as much photosynthesis as possible by positioning the foliage to get maximum sunlight—this is called canopy management. Then, in winter, you will prune the plant so that all that stored food is directed where you want it to go: into the shoots that will become strong canes with good foliage growth up top and healthy grapes along the fruiting zone. The goal is to have all canes be fruiting canes. Each cane will bear one to three bunches of grapes.

As you gain experience in your vineyard, you will learn how many canes to cut back, how many canes to leave as fruiting canes, and which canes to select from one year to the next. Avoid selecting "bull" canes—thick canes with widely spaced buds. But for those who have never pruned before, keep a few things in mind. Plants do not develop uniformly, so many of these rules do not apply 100 percent of the time. For instance, you may be unable to prune a cane to two buds because the spur is not mature enough and has only one bud. And remember that grapevines are vigorous and so forgiving plants. Don't hesitate to make a decision on pruning. Even if you make a mistake, the plant will grow again and give you another chance to get it right the following season.

Pruning is a once-per-year activity, done when the vines are dormant. However, in the first couple of years while you are training the vines on the trellis system, you may be shaping them with cuts during the growing season. This could be considered a form of pruning-training. I like to prune the dormant plants as late in the winter as possible; depending on your location, that would be from February to April. This late pruning avoids wintering of several diseases that can enter the vine through the pruning cuts.

PRUNING TIPS

Good pruning shears are valuable pieces of equipment. While there are many shears or clippers on the market, those that are inexpensive do not last long and cannot always make a sharp cut. So buy the best shears you can afford. Later, as the vines mature (around 10 years and up), you will need loppers to cut thicker canes.

Make cuts at a slight angle through the canes about a half-inch above the top bud. Cutting too close to the bud may damage it; cutting too high may leave a stub that dies back. The cuts may bleed a little sap, but that is not a problem.

After pruning, I leave the old canes that I have pruned on the vineyard floor between the rows. I let them dry out and then flail mow them into the vineyard or put them through a chipper. This way I put back into the ground what came out of the ground. The exception to this rule is if the canes have some kind of disease or insect that I want to eliminate from the vineyard.

Make clean pruning cuts with a sharp, sterile blade. Clean the pruners frequently. When canes start to reach a larger size, you may have to use loppers rather than hand pruners.

Training the Trunk

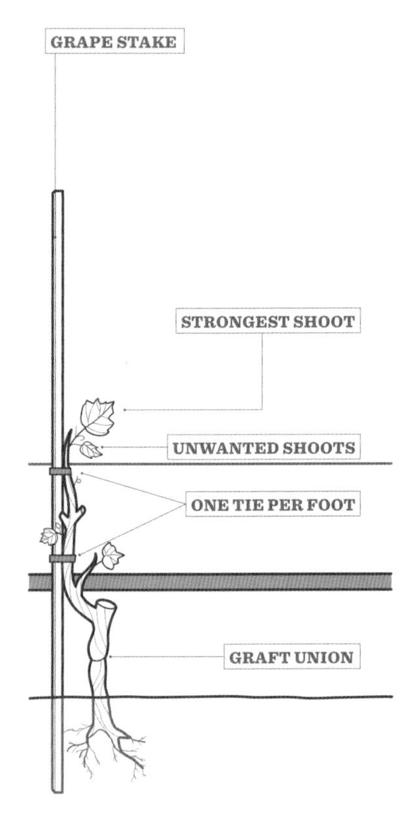

FIRST YEAR

The objective in the first year after planting is to produce a strong trunk, and cordons if possible. Root growth is rapid in the first year and then levels off to be consistent thereafter.

Training

As the vines grow, secure the strongest growing shoot of each vine with green plastic tape to its grape stake, about every 3 to 4 inches. Make sure that shoot has several leaves on it, as you will need these to form the cordons. This process develops the permanent trunk of your vine, so you want the trunk to grow straight and parallel with the stake. Once the strongest shoot reaches about 2 or 3 feet and you feel that it is healthy, pinch or prune off the other smaller shoots.

When at least two shoots of the vines grow above the fruiting wire by at least 6 inches, choose the shoot on the left closest to the fruiting wire and the shoot on the right closest to the fruiting wire. Tie these strong shoots to the wire; they will become your cordons or arms (the horizontal part of the vine). Once these shoots are clearly strong enough for you to see they will make good cordons, you can pinch or prune off the other vertical shoots. As the shoots continue to grow, keep tying each one to the fruiting wire until it reaches the center of the wire between the two grape stakes closest to the vine (this may not happen in the first year).

As each cordon begins to take definition (which will probably happen during the second year), you will select the strongest and most upright shoots along the cordon, about a hand's width apart. These will become your spurs.

Pinch off any little clusters of grapes in the first year so you direct the plant's energy into the roots, vines, cordons, and shoots.

Pruning

Your vines' leaves will start to turn yellow, red, and then brown in late fall. Don't worry, they haven't died—they are just taking a winter rest. They will be back in the spring. You will prune in late winter; January or February in warmer areas; later in cold-winter areas.

Your pruning this year will depend on how much of your trunk and cordons developed during the first growing season.

You will probably not have enough growth on the cordons in the first year to develop spurs. If a plant does have the basis for a spur, then skip ahead to instructions for second-year pruning for that plant. However, do not leave a spur on a cordon that has not developed a barklike surface. If the cordon does not have a barklike skin, prune off all spurs on the cordon.

If your vines just have cordons, prune back each cordon to the same length on each side of the trunk. If you have only one cordon, prune it off at the top of the grape stake. During the upcoming growing season you can develop both sides of the cordon. If your vine is very immature, be sure to prune back so you have at least two buds on the trunk, to leave two chances for continuing the growth of the trunk and then cordons in the months ahead.

The Spur

A spur is a cane (a one-year-old shoot) that has been pruned back to two buds.

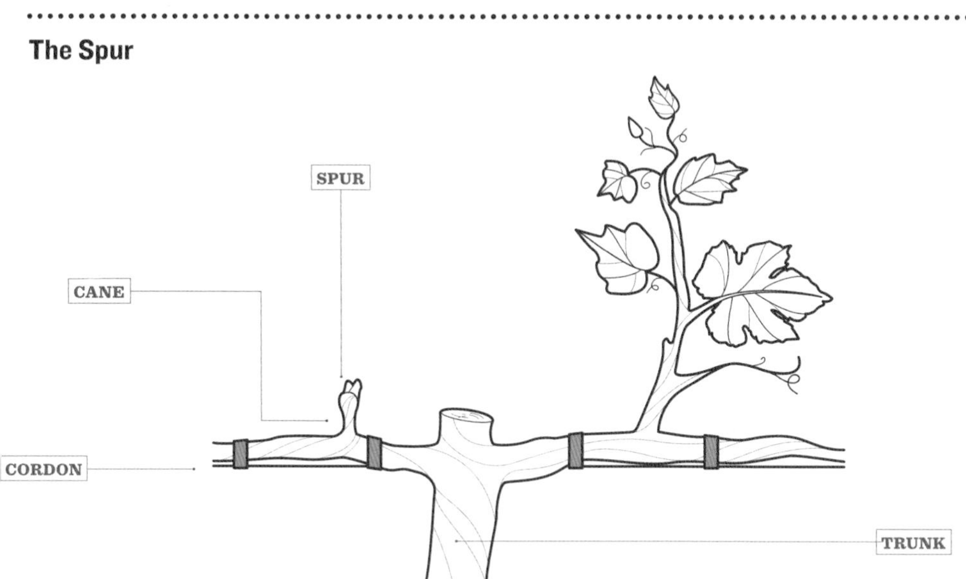

Pruning to Two Cordons

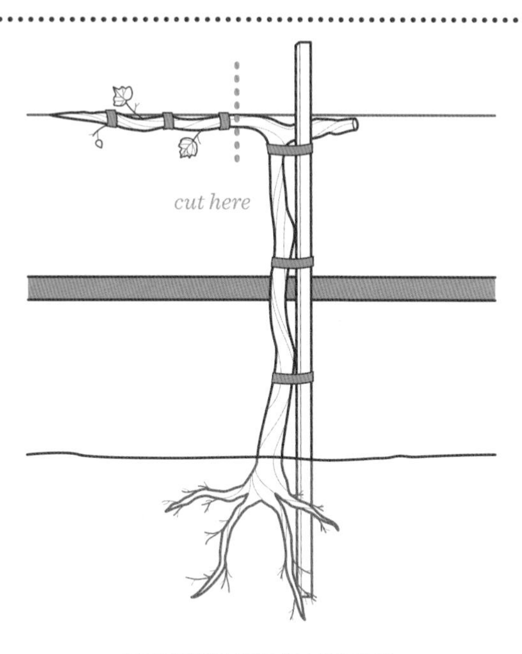

SECOND YEAR

The objective in the second year is to select and trim the shoots that grow out of the cordons. Ideally, you want to have 10 or 15 spurs on each cordon, with about 20 to 30 buds (or nodes), two on each spur. On some varieties, I like to leave fewer spurs, sometimes as few as seven or eight. That seems to improve the flavor of rich Bordeaux varieties (Cabernet Sauvignon, Merlot, Petit Verdot, Cabernet Franc, and Malbec). However, do not make this decision until the fifth year.

Training

In early spring, when the temperatures average 50°F (10°C), you will see the buds begin to push out. They will soon (at bud burst) be flowerlike, and then turn into green leaves. The leaves then grow and shoots will follow.

Continue tying vines up the stakes and out the fruiting wires. When each cordons reaches its desired length on the fruiting wire (the halfway point between one vine and the next), begin to concentrate on the shoots that come off of the cordon. Select about 10 to 20 shoots, each about a hand's width apart. These will become canes by the end of the year, at which point you will prune them to create the spurs of next year.

Thinning Shoots

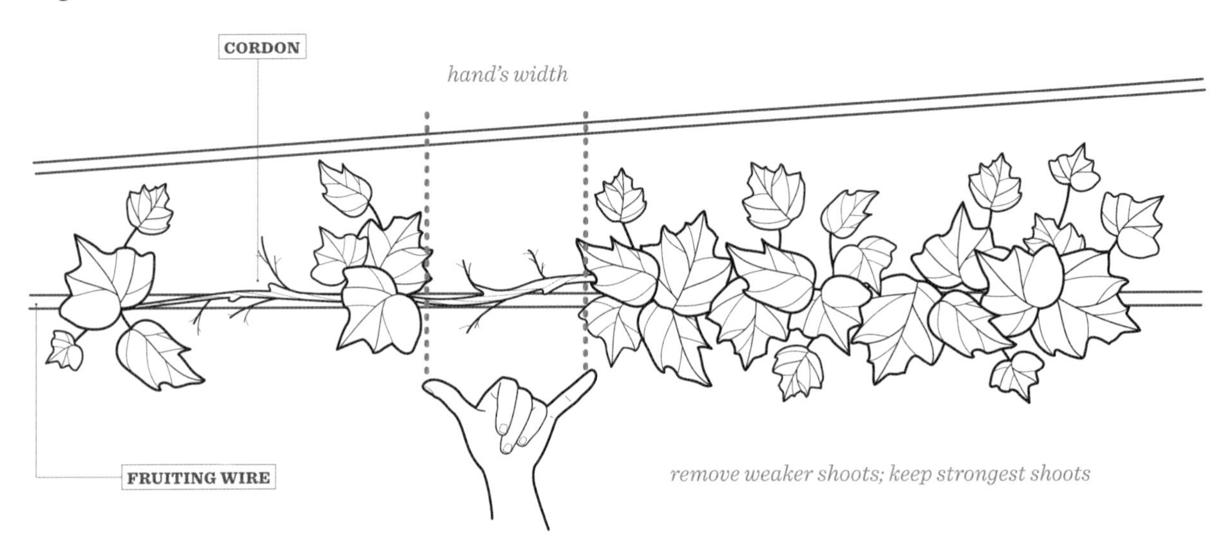

Try to choose those shoots that are the healthiest and strongest, that grow primarily upward, and that are relatively evenly spaced. Pinch off the rest of the shoots with your fingers. As the chosen canes grow up to the height of the training wires, push or gently place them between the double wires. Do not worry about mistakes, as vines are very forgiving and will grow back in a short time if you accidentally break them.

If you do not have material for 20 spurs, don't worry about it. Leave as many well-spaced spurs as you have material for. Also, because we are working with nature, perfectly even spacing or absolutely vertical upward growth is only a goal.

Pinch off the grapes this year as the trunk and cordon are not mature enough for a crop. The exception is if you planted in spring the previous year, in which case you may get a partial crop. If you have a cordon that has a barklike surface, you can allow the grapes to ripen.

Watch for and clip off suckers that come up from the base of the trunk or emerge from under the soil level.

Barklike wood on the trunk and the cordons is a sign of maturity in the grapevine. Don't let your vines bear grapes until the wood starts to develop some of this bark.

Pruning

Prune the vines in late winter, as before. The shoots that you selected during the growing season will now be canes. Prune the canes off to two buds (or nodes). These are your new spurs.

Move the positioning wires down closer to the fruiting wire so that you can train the canes as they start to grow in spring.

Pruning the First Spurs

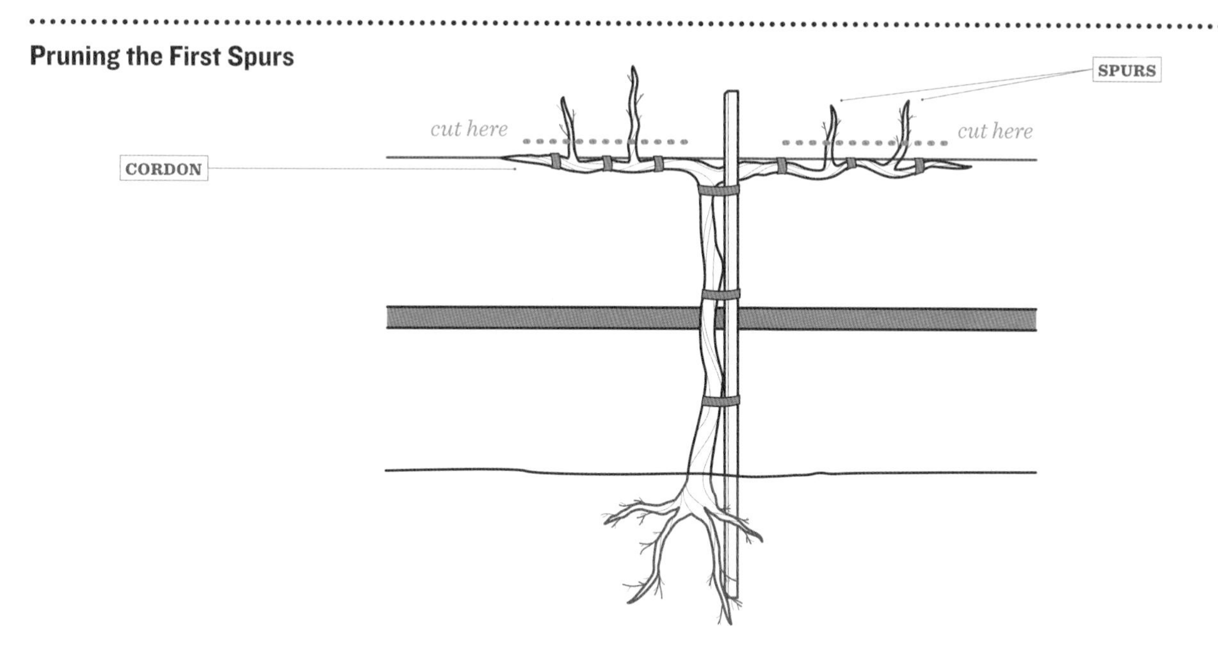

Green plastic tape is the best material for tying the cordons to the wire. It is flexible and won't damage the growing shoots, but does hold them firmly in place even when the canes start to produce fruit.

THIRD YEAR

In the third year the vines are continuing to consolidate root growth, and you are finally going for a small crop of grapes. You also want to continue to develop strong cordons on most of the vines and develop spurs on as many cordons as you can. The objective this year will be to build about 10 to 15 strong, well-spaced spurs on the cordon. The spacing began to be established last year, so you will try to add any additional spurs needed to fill in with shoots (now canes) from last year's spurs.

Training

Sometime around March or April, you will again see the buds begin to push out as they did in the second year. Follow the same training process with growing shoots as you did in prior years (tie them, pinch off additional shoots, and push the canes through the double positioning wires). Move the wires up the mid-posts as the canes grow.

Pinch off all the grapes on vines with smooth trunks or smooth cordons. Keep bunches on vines that have bark covering the main trunk and cordons, as these will be the only ones with grapes sweet enough to harvest. Generally, from 25 to 35 percent of the vines should have grapes that are good enough to harvest. If you have had two full years of vine growth, you may get more grapes, say 50 to 80 percent of vines. You will notice that the vines on the outer part of your vineyard (particularly in larger vineyards) will be more mature than the vines in the inner rows.

From early summer to late summer, you will see the shoots growing from the spurs into canes. Continue to push the canes between the training wires, moving the wires up as needed. Trim growth to 6 to 12 inches above the top training wires.

Pruning

Many of your vines will have well-established trunks and cordons with two canes each from last year's spurs. From each spur, choose the strongest and tallest of the two canes and prune it to two buds as the next spur. This is known as the renewal spur. Prune off the other cane.

CANOPY MANAGEMENT

As the canopy grows up in the third year and beyond, you will be removing laterals, shoots, and extra foliage to expose the fruit. Laterals are those shoots that grow horizontally from the main canes. A good time to start this is after the fruit has set, so that you don't disturb the process. If possible, time your disease management program so that you spray after you have pulled off the extra growth; this will allow you to get as much spray as possible into the exposed fruit.

In the hottest areas, concentrate on removing laterals first from the shady or "morning" side of the rows (that is, the side that gets light in the morning). Removing too much foliage from the "afternoon" side can lead to sunburn on the fruit.

Wear gloves to protect your hands. Work your way along the rows, pulling off any leaves around the fruit, as well as extra laterals. The objective is to leave enough foliage to provide dappled shade to the fruit, but to keep a shelf of protection over the fruiting zone. This typically means about 70 to 80 percent of the fruit is exposed. Gather up the foliage you have removed and add it to the compost pile.

These grapevines need some canopy management: pulling off the extra leaves, pushing the canopy up between the positioning wires, and exposing more of the fruit.

Removing Laterals

Work your way along each row pulling away the extra foliage from the fruiting zone and removing lateral shoots.

Renewal Spur

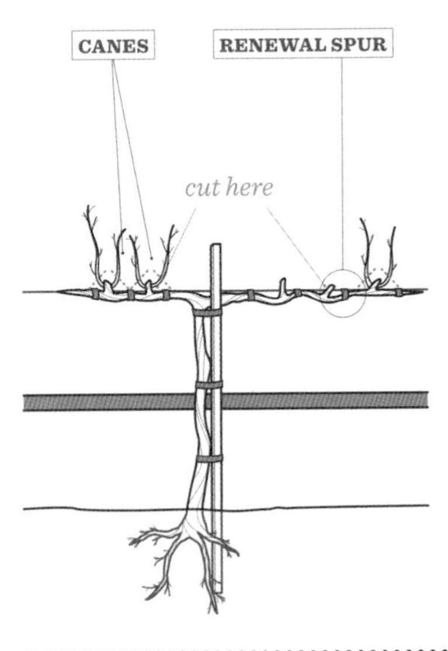

FOURTH YEAR

In the fourth year you are trying for a crop, with 70 to 90 percent of the vines having grapes. You may get more if you had a full three years of prior growth.

Training

Once again, the buds will push out in the spring and you will repeat the basic maintenance steps of last year.

In late spring, pinch off the top bunch of grapes at flowering on those vines that are mature (have a barklike skin on the trunk and cordon). On immature vines (those without bark) pinch off all bunches of grapes. The exception to this is if the vine trunk is large; then you can leave all grapes on.

Start canopy management, shoot trimming, and leaf pulling. During summer, if your vines have vigorous growth (lots of leaves, long internodes, and lateral shoots),

Trimming the Top Growth

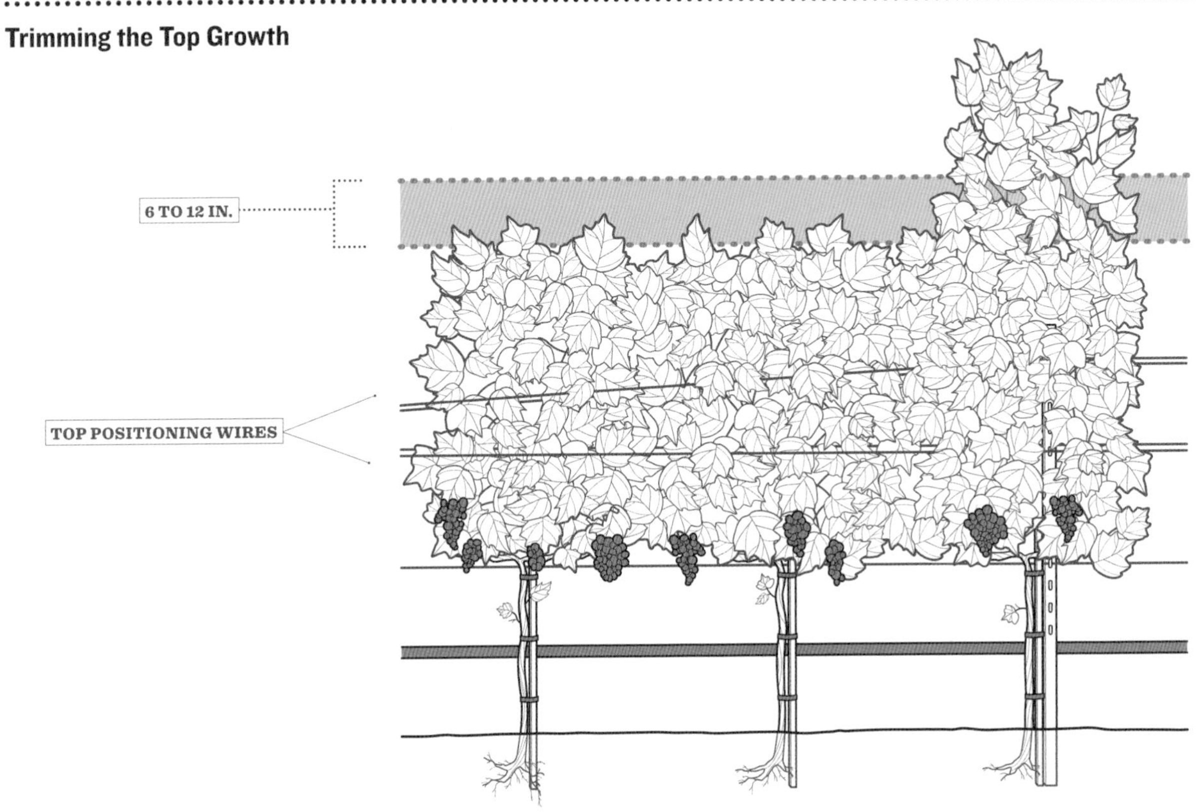

pull off more leaves to allow the grapes to get some direct sun. About 70 to 80 percent of the grapes should be exposed to the sun. If your vines grow numerous lateral shoots, you are watering too much.

You should cut the shoots at the top again to about 12 inches above the top positioning wires.

By the time you are getting close to harvest, you want the fruit to be exposed and extra laterals and low-growing leaves removed.

Pruning

This year, the objective is to fill out the vine with about 10 to 15 strong spurs, each with two buds, for a total of 20 to 30 buds per plant. This is just a guide—you will want to vary the number of buds: fewer if the vine is very vigorous; more if the vine is less vigorous.

For example, if the plant had weak growth during the growing season, try pruning every other spur to one bud and the remaining spur to two buds. Remember, most of our new spurs, this year and hereafter, will be renewal spurs.

If the plant was excessively vigorous during the growing season (too many leaves, long internodes, and many lateral shoots), prune to increase the number of buds on the spurs to create more shoots. For instance, try pruning every other spur to three buds and the remaining spur to two buds.

Well-balanced Pruning

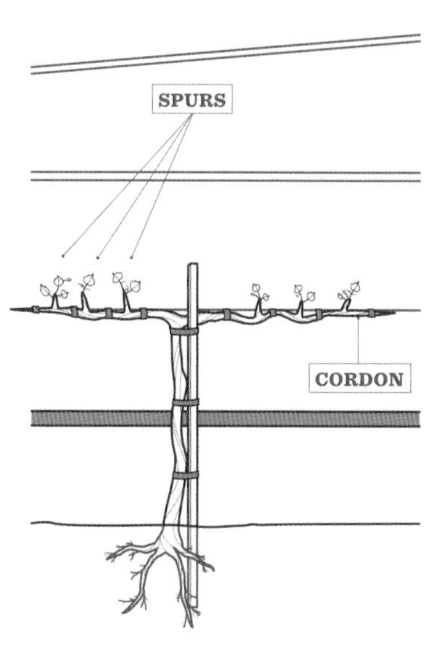

On mature vines, you can see the spurs that have been carefully selected to serve as renewal spurs during each year's dormant pruning.

FIFTH YEAR AND BEYOND

The objective in the fifth and following years is to have well-balanced plants, each with the correct number of spurs to produce healthy fruit that is evenly distributed along the cordons.

Training

Continue to manage the canopy and yield. On weaker plants, pinch off the top bunch of grapes when they are flowering. Tie and position shoots that will become this year's canes. Trim top shoots and pull leaves as in prior years. If you are growing the Bordeaux varietals mentioned earlier—Cabernet Sauvignon, Merlot, Petit Verdot, Cabernet Franc, and Malbec—this is the year you will consider removing some spurs to get them down to seven or eight maximum.

Pruning

A mature plant allows you to perfect your pruning process, year after year. One mistake I see is to allow the spurs to get too large, one on top of the other. When pruning, try to pick the canes that are closest to the cordon.

MANAGING **EXCESS VIGOR**

The primary reason you may want to modify your trellis system is that your vines are growing too well. That is, they are too vigorous. This is usually because of choosing an overly vigorous rootstock or variety, or because you have very rich soil. It can also be a consequence of overwatering or fertilizing. You can either try to reduce the vigor of the vines or create additional space for those extra leaf surfaces.

There are four basic methods to manage vigor for optimum photosynthesis:

- Pull more leaves, trim lateral shoots more, and trim the tops of the shoots more often—this is an ongoing need that will mean more labor, so permanent fixes may be better.
- Pinch off the top cluster of grapes before flowering.
- When you prune in late winter, on the most vigorous vines, allow three buds on every other spur; also, on the weak vines, consider allowing only one bud on every other spur alternating with two buds on the adjacent spur.
- Finally, consider developing additional spurs with downward pointing shoots, and training existing spurs downward. For this procedure, you may have to add another wire or a set of double wires, 12 to 14 in. below the fruiting wire, to position the downward-growing shoots.

A Modified Trellis

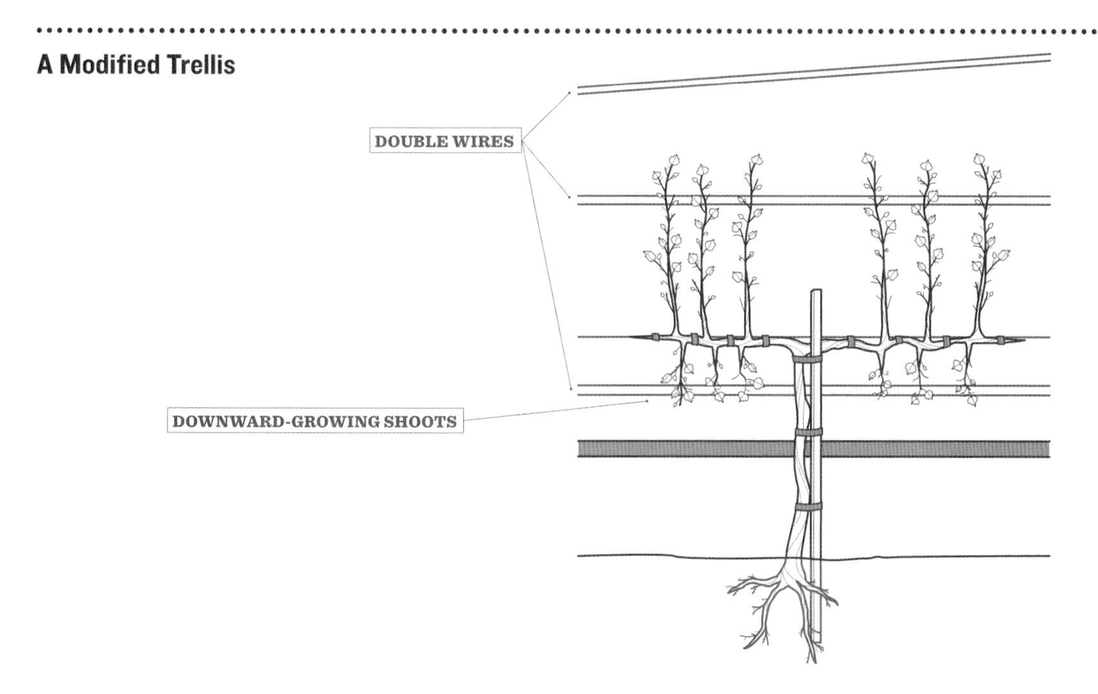

MAINTENANCE THROUGH THE YEAR

IRRIGATION

FERTILIZING

Maintaining THE VINEYARD

ORGANIC FERTILIZERS

WEEDING AND COVER CROPS

MAINTENANCE THROUGH THE YEAR

In addition to pruning and training your vines, other ongoing tasks required to take care of your vineyard include:

- Watering correctly.
- Addressing any nutrient deficiencies.
- Controlling weeds and otherwise maintaining a good ground cover or cover crop.
- Managing and treating fungal diseases.
- Deterring vineyard pests, such as deer and other grazing animals, as well as insects and soil pathogens.

IRRIGATION

During the first six weeks after planting, the vines need 1 gallon of water per day until the roots are well set. After that, they need less and less water. Back off to every other day for three weeks, then every third day for the rest of the first year. Calculate the amount of water that the vines are getting based on your rainfall (if any), the emitter rate, and your soil type. For example, if you have 1/2 gallon per hour (gph) drippers, you will turn on the system for two hours to deliver 1 gal. of water. Use some judgment here and make sure that the weather is warm enough to dry out the ground around the drippers enough so the plants' roots are not always in wet soil. If you have a tensiometer, use it to evaluate the soil moisture.

It is not necessary to water after the plants have become dormant. However, it is appropriate to water in the first and second year after leaf fall. A second root growth begins in the fall and continues for several weeks into the winter. So after the leaves have completely fallen, if the soil is still dry, you should give a good dose of water to your vineyard, to allow this root growth to flourish.

Second Year

What you are trying to accomplish in the second year (and the third year and thereafter), is to have your vines' roots find, or struggle to find, their own water source in the ground. You do this by being somewhat stingy with the water, but not so stingy that the shoot ends do not grow, or the leaves look gray and the plant generally does not look healthy. Likewise, you do not want excessive shoot growth, which is recognized by large leaves, long internodes, and excessive lateral shoot growth.

Grapevines are pretty drought-resistant, but if you do not give them enough water, they will soon show signs of it. It's better to under-water your vines than to water them too much. But don't wait long; if you're concerned, keep a daily eye on the vines for signs of water stress on the shoots. In the second year of growth, make sure that the vines' roots are not sitting in wet soil.

Root growth and shoot growth are interrelated. Remember that your plants have had two seasons of root growth. Depending on when you planted last year, your roots have grown down 4 to 8 inches, possibly more, into the soil down so they are firmly set in the ground and pulling their own moisture from the bottom of their little roots.

In medium heat areas with low summer rain, water the vines every third day for one hour in early to midsummer. If it gets really hot and dry for three or more days, give the vines a deeper watering, which could be two hours instead of one. After midsummer, water every five days for two hours.

Again, depending on your climate and soil, use some judgment about how

IRRIGATION SCHEDULE YEAR BY YEAR

	SOIL TYPE		
	SAND	**LOAM**	**CLAY**
FIRST YEAR			
Weeks 1 to 6 after planting	1 1/2 gal. per day	1 gal. per day	3/4 gal. per day
Weeks 7, 8, and 9 after planting	1 1/2 gal. every second day	1 gal. every second day	3/4 gal. every second day
Remainder of first growing season	1 1/2 gal. every third day	1 gal. every third day	3/4 gal. every third day
SECOND YEAR			
Early to midsummer	1 1/2 gal. every third day	1 gal. every third day	3/4 gal. every third day
Midsummer to midfall	2 1/2 gal. every fifth day	2 gal. every fifth day	1 1/2 gal. every second day
THIRD YEAR			
Late spring to early summer (until veraison)	4 gal. every 5 to 7 days	3 gal. every 5 to 7 days	2 gal. every 5 to 7 days
FOURTH YEAR			
Late spring to early summer (until veraison)	5 to 7 gal. after hot spell	3 to 5 gal. after hot spell	2 to 3 gal. after hot spell

much to water. In hotter and drier climates, start irrigating two to three weeks earlier and increase the amount of water by 25 to 40 percent. In cooler climates, start two to three weeks later and decrease the amount of water by 25 to 40 percent. Naturally you will not water during rainy periods and at leaf fall.

Third Year

In the third year, do not water until it starts to get warm and the soil dries out; say late spring or early summer, depending on your climate and soil type. You may want to switch your drippers from 1/2 gph to 1 gph. Water deeply for six hours with 1/2 gph or three hours with 1 gph drippers every five to seven days, depending on hot spells. If temperatures are cool or if it is a wet year, you will water less. The hotter it is, the more frequently you should water. Keep an eye on the shoot growth and adjust accordingly.

Begin to cut back water after veraison unless you see severe signs of drought. You will surely have a harvest this year. As soon as you have harvested the ripe grapes, give the plants an ample watering; say, four to eight hours. Then resume your watering schedule. Again, cease all watering during rainy periods and after leaf fall.

Fourth Year and After

In the fourth year and thereafter, you will have good root growth and the vines are getting water on their own most of the time. You should water during long spells of very hot weather, and then give a deep watering for three to five hours. Continue

Shoots that start to shrivel at the tips are a sign that the vines are not getting enough water.

Young grapes like these need more supplemental water than they will when they become mature. Mulches and cover crops help to retain water in the soil.

to keep an eye on the shoots to see if there are signs of water deprivation. Begin to cut back on water after veraison, unless there are signs of drought or extreme water deprivation.

After your harvest, give the vines four to eight hours of water and resume your water schedule until it rains or leaf fall, then stop watering.

Many mature vineyards are not irrigated except in drought years, and vineyards in some regions receive all the water they need from rainfall during the growing season. Once your vineyard is mature, irrigate it according to the water supply and vine needs. But no matter what, always cut back with approaching harvest and again with vine dormancy to assure fruit and wood maturity.

FERTILIZING

If your pre-planting soil test showed any major soil nutrient deficiencies, you will have amended the soil before putting in the vineyard. You may need to do this again if future soil tests show major deficiencies. As your grapes grow, you'll need to ensure they get the nutrition they need on an ongoing basis. But unlike many other crops, grapevines are not heavy feeders. In fact, too much fertilizer will encourage excessive leaf and shoot growth.

If you suspect that your vines have nutrient deficiencies, you can once again have a soil sample performed by a laboratory. Another form of testing that is even more precise for an established vineyard is a petiole test. A petiole is the stem that attaches the leaf to the shoot or cane. It's not necessary to do a petiole test on a regular basis in a home vineyard, but it can help you identify specific nutrient deficiencies.

Petiole analysis is best done after bloom, but a testing service in your area will advise you of the precise timing based on common deficiencies in your region. If you have observed symptoms in some plants and not others, collect samples from both healthy and affected vines. If you grow more than one variety, sample each variety separately. Choose petioles from mature leaves five to seven leaves from the tip of the shoot. Petiole analysis is usually done for nitrogen, phosphorus, potassium, boron, calcium, copper, iron, magnesium, manganese, and zinc.

Major nutrient fertilizers can be applied in several ways. You can sprinkle powders or pellets directly on the ground under and around the plant; if you have a drip system, apply them under each dripper. Other fertilizers are soluble, and are mixed with water and applied through the irrigation system: this normally requires you to have a tank and pump to disperse the product into the system. If you apply foliar sprays with a backpack sprayer, be sure to use dedicated spray equipment that has not been used for a pesticide or fungicide.

Fertilize once when plants are dormant, taking care not to apply the fertilizer too early, because that will stimulate new shoot growth. Early winter is the best time: you want the plants to be dormant, so that the fertilizer doesn't cause a flush of late growth that could be subject to frosts, but the winter rains will wash the fertilizer down into the root zone in preparation for spring growth.

If you have had a soil or leaf (petiole) sample that showed deficiencies, recommended application rates will be included in the analysis. Otherwise, use a balanced fertilizer according to the rates specified by the manufacturer or supplier. If your soil is very fertile, cut the suggested amount in half or even to a quarter, likewise for clay soils, which retain nutrients. In sandy soils, double the amount of fertilizer, as nutrients will leach more quickly out of the plants' root zones.

Fertilize the vineyard again in early spring, generally basing the vines' needs on the amount of vigor they show. Apply after heavy winter rains, but early enough so that spring rains can move nitrogen into the root zone (especially if you have an unirrigated vineyard). Late spring and early summer applications at light rates are recommended for drip-irrigated vineyards, again, only if the plants lack vigor.

Compost for the Vineyard

Composting is one way to minimize the amount of additional nutrients you need to supply to the vineyard. Creating your own compost pile or piles is almost essential for an organic farmer. Compost also helps to improve soil texture, by adding organic matter to the soil and increasing soil microbial activity.

There will always be some material around to dispose of. Kitchen scraps, grass clippings, cuttings from the vines, garden waste—anything that can be broken down can be added to the compost. If you are making your own wine, grape pomace is a great addition to the compost pile.

Never put into the compost any material that has been infected with a disease or pest, however, as it can survive the composting process to reinfect the vineyard. Do not add annual weeds that have gone to seed, or even roots of persistent perennial weeds such as blackberry, which can be surprisingly resilient.

A good method of composting is to have a two- or even three-bin system. As the compost matures, you move it into the adjacent bin. This also helps to aerate the compost, which is necessary to improve the aerobic process of breaking down the waste. Bigger piles can get quite hot in the interior; temperatures of 130°F (55°C) or more are needed to break down the material.

The best time to apply compost is in fall, before the ground freezes. Spread it under the trellis rows up to a thickness of several inches.

ORGANIC **FERTILIZERS**

As with other organic products, there are different opinions about what is considered an organic fertilizer. A good rule of thumb is that an organic fertilizer is one that is non-synthesized, that is, it comes from a source found in nature. This can mean everything from fish to cow bones to earthworm castings. But it does not mean that organic fertilizers only come from living sources. Minerals are also a major component of organic fertilizers.

One of the main differences between organic and non-organic fertilizers is that most organic fertilizers release their nutrients more slowly and, sometimes, unpredictably. This can be beneficial in the vineyard unless your vines are suffering from serious deficiencies. Sudden burst of growth fueled by fertilizers are subject to predation by pests. Grapevines are typically vigorous enough to grow well without such major intervention.

In my vineyard, I prefer to use compost from my own vineyards, chicken and turkey manure, fish and bone meal, and manure from my own sheep, who graze in the vineyards when the vines are dormant. The list provided here contains examples of some of the organic materials you can consider. Some of these, such as bat guano, can be expensive. Others may be available in limited quantities, or not readily available in your area. Shop around, and be sure to develop your own sources of compost. I have learned a lot from the Organic Fertilizer Association of California. I suggest you turn to a similar organization in your area if you have questions or concerns about fertilizers in your vineyard.

Fertilizer analysis is written as a percentage of N-P-K for the major nutrients. The list here gives typical values for a range of organic fertilizers. Packaged commercial formulas will also list any minor nutrients, such as boron or magnesium.

- alfalfa meal (4-1-1)
- blood meal (12-1-0.6)
- bone meal (4-12-0)
- corn gluten (10-0-0)
- cottonseed meal (6-1-1)
- feather meal (14-0-0)
- fish emulsion (5-2-2)
- fish meal (10-6-2)
- greensand (glauconite) (0-0-5)
- guano (bat) (11-8-2)
- guano (bird) (13-12-3
- kelp (1-0-3)
- peanut meal (6-1-1)
- potassium chloride, muriate of potash (0-0-60)
- potassium sulfate, sulfate of potash (0-0-50)
- rock phosphate (0-3-0)
- soybean meal (7-1-5)
- sodium nitrate, nitrate of soda (16-0-0) Permitted for limited use in organic farming.
- pelleted chicken or turkey manure (1-1-0.5)
- urea (60-0-0)
- wood ash (0-2-5) Note that only wood ash is considered organic. Don't use ash from other sources.
- worm castings (2-1-1)

FEEDING YOUR VINES

Nitrogen (N)

Nitrogen is usually the major nutrient most needed by grapevines. When nitrogen is low, there is reduced shoot growth, which leads in turn to a reduced canopy and lower photosynthesis—eventually resulting in lower yields. Excess nitrogen causes too much vegetative growth, which leads to overshading by adjacent vines and reduced fruit quality.

The best time to apply nitrogen is during periods of root growth (from fall to early winter and spring—but before veraison). I find that in the spring around fruit set is the best time. Ammonium nitrate and urea are traditional agricultural sources of nitrogen but these do not always come from an organic source. Alternative organic sources of nitrogen include alfalfa meal, cottonseed meal, soybean meal, nitrate of soda (sodium nitrate), earthworm castings, and bat guano. You can also add nitrogen to the soil by planting legumes as cover crops (clover, peas, fava, and other beans), and by developing your own compost.

Phosphorus (P)

Phosphorus promotes root development, bloom, and strong structures (trunk and cordons). Common sources of phosphorus are rock phosphate, bat guano, bone meal, and fish meal. You can usually give an annual application of phosphorus along with nitrogen and potassium. Phosphorus deficiency is harder to detect visually, but can be demonstrated by a weak bloom and spindly structure.

Potassium (K)

Potassium promotes strong flowering, fruit production, and general health and resistance to pests. Sources of potassium are sulfate of potash, muriate of potash, wood ash, manures (such as chicken, turkey, or horse), and potassium nitrate. Potassium deficiencies show up in midsummer to late summer with bronze-colored leaves. Some leaves may have dark reddish spots, margins, blotches, or dead spots.

If you continue to build your soil and feed your vines, they are unlikely to suffer from nutrient deficiencies. Be sure to identify any nutrient problems through a petiole test. This leaf is showing signs of potassium deficiency.

WEEDING AND COVER CROPS

Weeds in the vineyard are one of the biggest challenges for the organic grower. They use up sunlight, nutrients, and water that you would prefer to have going to your grapevines. Weed competition in the first year can reduce vine growth by as much as 50 percent; in worst cases, they can choke your vines. Weeds are hosts for pests and for fungal diseases. Those with large, succulent roots can attract gophers.

If you have prepared your site well, you will hopefully have cleared the worst of the perennial weeds. But you will never be free of annual weeds and so controlling them is an ongoing maintenance task in the vineyard. Cover crops planted in the middle of the rows and physical removal of weeds at the base of vines under the trellis are the basis of your weed management program.

At the base of the vines, keep the weeds down by clearing weeds with a string trimmer, shovel, hoe, mulch, or an organic weed control spray. I have used a spray with 25 percent vinegar with some success. Other organic contact herbicides may be soap-based or contain some other material, such as clove or citrus oil. Be careful not to spray any herbicide on the developing vine; in the first year, your grow tubes will protect the young vines but after that you may have to use primarily mechanical means of weed control close to the trunk. If you do have grow tubes in place, you may need to lift the tubes to clear out any weeds growing up inside them.

Some growers use a thick mulch around the vines and under the wires to help keep down weeds. This can work well in a small vineyard, but you will have to replenish the mulch every one to three years. Other growers use a small propane flame thrower to control weeds. I prefer not to use these devices as I am concerned about hurting my vines and potentially causing a fire.

Between rows, mowing and cover crops are the best methods of weed control. Mowing provides mulch—as long as you leave the clippings on the ground—and a natural source of nutrients. You can also use a small riding mower or even a handheld string trimmer to break up any larger prunings on the ground.

Manure and compost are great amendments for building soil structure and fertility. If you get fresh manure from any source, be sure to compost or age it before application. Also make sure it is weed-free.

In mild-winter climates, a cover crop can grow through the winter months even when the vines are dormant. In spring, you can mow or disc it into the soil, where it will add fertility and structure.

Cover Crops

Many vineyards maintain a cover crop or ground cover year-round between the rows, especially in mild climates where the ground cover can survive without supplemental irrigation. In other vineyards, cover crops are planted in fall or spring and then mown or disced into ground in late spring or early summer.

Cover crops have many benefits in addition to weed control.

- They can control soil erosion, particularly in hillside vineyards.
- They help with frost control.
- They act as a mulch to retain soil moisture and to keep down dust.
- They supply a natural source of nitrogen and other nutrients. They may also increase the availability of nutrients.
- They provide a habitat for beneficial insects.
- They increase biological diversity in the top layers of soil, by encouraging earthworms and other beneficial soil-dwelling insects.

→ They improve soil structure and water-holding capacity, as well as improve drainage and prevent soil compaction.

Organic fertilizers are substances that are not manufactured by a chemical process. They come from natural sources, either animal, plant, or mineral.

Late fall and early winter is the best time to plant a seasonal cover crop or ground cover between the rows, so that fall and winter rains can help them to grow. High-nitrogen cover crops such as fava beans especially benefit during the rainy season in mild-winter climates. Some cover crops can be planted in the spring, or you may need to plant in spring if you live in a very-cold-winter region.

A wide variety of plants can be used as cover crops, such as clover, vetch, fescue, or ryegrass. You can buy specialty blends, such as all legumes or mixes to attract beneficial insects. For even coverage, buy a seed dispenser to spread the seed between the rows of your vineyard. Do not allow the cover crop to grow under the vines; allow a 14 to 18 inch strip of bare ground under the trellis. In the first year, do not let the cover crop grow too high; mow it after 3 to 4 inches of growth.

When you mow the cover crop, leave the mowings/clippings on the ground. Keep in mind that tall cover crops like legumes and especially mustard can be hard to cut down by hand, so unless you have a decent mower, stick to clovers or other lower-growing cover crops.

To encourage seeding for the next year's cover crop, allow the plants to go to seed and then rototill them into the ground or mow the mature crop. If you collect seeds from the previous year's cover crop or buy seed, keep them in a paper bag in a cool, dry place until you are ready to sow.

COVER CROPS

COVER CROP	SOIL TYPE	SEASON	COMMENTS
Crimson clover	sandy to loam	winter-spring annual	Taller growth (12 to 18 in.) may interfere with frost protection; mow at 3 to 4 in. Attracts bees.
Fava beans	loam	winter-spring annual	May need to be treated with an inoculant before planting. Does not self-sow. Subject to black aphids, but they will not affect grapevines. Produces large pods of edible beans. Does not tolerate close mowing; allow to grow to full height before discing in.
Lana vetch	sandy loam to clay	spring annual	Earliest maturing vetch; can produce a dense, competitive, viney stand. Establishes easily and reseeds well. Mow at 5 in. and allow 4 weeks regrowth before maturity for seed production.
Medics, including burclover	sandy to clay	winter-spring annual	Drought-tolerant; most stands of burclover are volunteers. Growth is competitive, especially where mowing heights of 3 to 5 in. restrict growth of other plants.
Rose clover	sandy to loam	spring annual	Early maturing; drought tolerant. Mow early at 2 to 4 in. to control competing vegetation. Allow 3 to 4 weeks regrowth to set and mature seed. Reliable reseeder. Good for erosion control.
Subclover	sandy loam to clay	winter-spring annual	Early maturing with prostrate growth habit. Keep down competing weeds during establishment. Best adapted to soils with good water-holding capacity.

COVER CROPS

COVER CROP	SOIL TYPE	SEASON	COMMENTS
White vetch	loam	spring annual	Produces a dense, competitive stand. Mow at 5 in. and allow 4 weeks regrowth before maturity for seed production.
Winter peas	sandy to loam	winter annual	Many kinds available; check with local sources. Also used as a forage crop. Adds nitrogen to soil and improves soil texture.
GRASSES			
Annual bluegrass	most soils	winter annual	Low growth habit and shallow-rooted; does not compete with other plants. Best used where frequent close mowing is practiced. Early development and maturity make it useful for erosion control.
Annual ryegrass	loam to clay	winter annual	Competitive plants, with erect, leafy, vigorous growth and a heavy fibrous root system. May cause pollen allergies.
Barley	most soils	winter annual	Fibrous root system reaches deep into soil; grows quickly. Does not overwinter in coldest areas. Good for controlling weeds.
Blando bromegrass	most soils	winter annual	Reseeds reliably. Short stature with semi-erect, vigorous growth. Mow early at 2 to 3 in. for frost protection; discontinue 3 to 4 weeks before maturity.

COVER CROPS *continued*

COVER CROP	SOIL TYPE	SEASON	COMMENTS
Triticale	most soils	winter annual	Winter-hardy wheat relative. Deep root penetration; good for reducing soil erosion. Can also be sown in spring as ground cover.
'Zorro' annual fescue	most soils	winter annual	Drought tolerant, early maturing with good reseeding ability. Manage as bromegrass, except mow to 4 in.

Clover *Triticale* *Vetch*

INTEGRATED VINEYARD MANAGEMENT

SAFE USE AND *DISPOSAL OF* PESTICIDES

ORGANIC DISEASE CONTROL

DISEASES AND PESTS

GRAPEVINE DISEASES

VINEYARD PESTS

INSECTS IN THE VINEYARDS

If you are tempted to spot-treat weeds with a propane flamer, exercise extreme caution. It is easy to damage your grapevines or even start a fire.

INTEGRATED VINEYARD MANAGEMENT

In order to manage your vineyard in a sustainable fashion, you will use a variety of approaches to deal with pests and diseases. It is not enough simply to spray a chemical on the problem and hope it will go away. Remember, you are trying to work with nature rather than against it, using the resources at hand to help manage problems such as pests and diseases.

The first step is to practice prevention. Correct vineyard layout, good soil preparation, and selecting the right rootstock will help to ensure that your grapevines can resist many problems. Next is maintenance. Plants that are weak or stressed through lack of water or nutrition are more subject to predation by insects. Overcrowded vines with too much foliage are more likely to develop fungal diseases due to lack of circulation. Weeds compete with vines for resources and they also harbor diseases and pests. So keeping up with your maintenance tasks through the year—watering, feeding, canopy management, and weed control—helps to keep down the number of problems you have to deal with.

But inevitably diseases and pests will find your vines. When this happens, you want to be sure which pests you have. Dealing with problems in the vineyard is a process of four steps.

1. Observation

Your initial pest and disease control program should start with an ongoing visual observation of the vineyard. Walk through the rows and look for anything unusual or that has changed significantly. In other words, know your vineyard. When you first discover something you think is different, decide whether it is within the normal range or whether it looks like a problem.

2. Identification

Consider what the elements of the problem are. Look at the growth. Is it weak or excessive? Are shoots curling? Are leaves eaten or scarred? Is the color changed? Is there a distortion of the leaf surface? Are insects present? Observe not only the leaves, grapes, and shoots, but the trunk and roots. Snoop around and under the leaves, peel off some bark, and pick off a grape. Is it only one vine? Is the problem localized or is it scattered throughout the vineyard?

3. Diagnosis

After you have carefully observed the affected vines and compared them to healthy, normal vines, you will begin to determine what is eating or damaging your vines. A number of common problems are listed in the following pages. But if you cannot positively identify a pest or disease, don't guess at it. Talk to other local growers and farmers. Consult your local extension service or agriculture office. Consult a reliable online source, such as that of your local extension, or a good national source like Cornell University or University of California, Davis. An excellent printed resource is the UC Davis book *Grape Pest Management.* It is very helpful, covers all aspects of pest management, and has great photos to help you compare what may be happening to your vines.

If all that fails, I recommend that you take a sample of the leaf, shoot, or other part of the vine, and send it to a commercial agricultural testing laboratory for evaluation. Many services will tell you how to package a sample, how to send it, and how much it will cost for the evaluation and recommended treatment.

A cover crop or ground cover of grasses and clovers has many benefits. It attracts beneficial insects, helps keep down the weeds, and helps prevent erosion of the soil.

4. Treatment

Always choose the least-toxic method of control for any problem. Practice prevention and good vineyard hygiene, continue to build healthy soil, and encourage beneficial organisms. You will find that most problems can be dealt with.

Don't overreact or panic when you find a pest in your vineyard. I have found white flies, large worms, and discoloration or loss of leaves, and determined after research they were indeed pests but not significant enough to worry about. Organic vineyards never look perfect because some insects, both beneficial and otherwise, remain in the vineyard.

SAFE USE AND DISPOSAL OF **PESTICIDES**

I believe that most problems in the vineyard can be handled with organic products. But just because a product is in a green bottle doesn't mean that it is acceptable under organic standards. These standards differ depending on the state or province where you live, but all the products I am going to recommend are acceptable under the USDA National Organic Program (NOP) classification system. You will find contact information for agencies and organizations in Resources.

Even organic products call for following some sensible guidelines.

- → Use of the least-toxic treatment products is always the best approach.
- → Make sure the product you choose is labeled for use on that plant and for that disease or pest.
- → Read the entire label—it's the best source of information about how to use the product. If the label does not answer your questions, ask for the manufacturer's fact sheet, or look them up online.
- → Time your treatment to catch pests or diseases at their most vulnerable stage.
- → Buy only the amount you can use in one season.
- → Spot treat whenever possible.
- → If you buy concentrates and dilute them for use, mix up the smallest quantity that will do the job, and use it up.
- → When mixing or measuring products, use a dedicated set of spoons and cups. Mark the utensils with something permanent, such as fingernail polish, and store them with the products so they won't ever be used in the kitchen.
- → Store products in their original containers, out of the reach of children and pets, in a cool, dark, dry place outside the house. A locked cupboard in the garage or storage shed is ideal.
- → Never remove labels or transfer products to other than their original containers.
- → Water used to rinse out a sprayer or applicator should be applied like the pesticide.
- → Some states, provinces, or counties, such as California, regulate the use of chemicals, so you may need to obtain a license from your county or state agriculture department; even organic products may be covered under licensing requirements. These licenses for home treatment are usually very easy to obtain. While organic products are generally of lower toxicity than those not rated organic, they are still not safe to drink or eat—after all, they are

used as pest control because they have the ability to harm living organisms. That's why safe handling, storage, and disposal practices are necessary, and training and licensing may be required. Even where there are no licensing requirements, strict adherence to label directions, handling, application, and disposal is important.

ORGANIC DISEASE CONTROL

There are very few vineyards, large or small, that do not need to spray to control fungal diseases. If you have chosen the best variety and rootstock for your site, and you practice good canopy management and keep the vineyard clean, you will be able to prevent some fungal disease problems, but you are unlikely to eliminate the need for a spray program entirely. Fortunately, there are quite a few products that are considered safe for use in organic vineyards. Most organic fungicides are based on the traditional antifungal agents of sulfur and copper in different formulations. Bordeaux mixture, for example, is a sulfur and copper spray that has been used for many years in vineyards. (Because copper and sulfur are elemental, fungi do not develop resistance to them.) You will find these products in various forms, typically powdered, liquid, or granules. Sometimes the active ingredient is combined with other ingredients, such as oils or spreader/stickers. These may help the active ingredient to adhere to the grapevines, or may have some antifungal action of their own.

In general, waiting to see mildew spores on your leaves means that you have left it too late to start spraying that year. Fungal spores almost always overwinter

Bordeaux mixture is a traditional spray used to control fungal diseases. It is used in organic vineyards around the world.

on the ground or in the plants and then are released by spring rains and rising temperatures, and their growth is often encouraged by humidity. The time to start most spraying programs is before or around bloom, and then you must continue to reapply every week or two. You may wish to try different products and to alternate products. As your vineyard grows, you will settle on a schedule for spraying in response to your experience, and then vary it according to the conditions.

There are also some bacterial and viral diseases that affect grapevines. Although bacterial diseases can be controlled with various measures, your best defense against viral diseases is to buy only certified virus-free stock. Controlling pests that spread diseases—such as nematodes and mealybugs—will also help to prevent transmission of viruses. Virus-infected plants cannot be cured.

The National Organic Program list titled Allowed and Prohibited Substances is a good overall guide to disease controls, although states and provinces may have their own specifications. Here is a list of substances generally used to treat diseases in organic vineyards. You will often find one or more of these in commercial preparations labeled for organic use:

Aqueous potassium silicate
Bordeaux mixture
Coppers, fixed: copper hydroxide, copper oxide, copper oxychloride
Copper sulfate
Elemental sulfur
Hydrated lime
Hydrogen peroxide
Lime sulfur
Horticultural oils
Potassium bicarbonate

Using a Backpack Sprayer

A backpack sprayer is the easiest way to apply your fungal spray if you have more than a small row of vines. It's easy on the back and the long wand allows you to reach all of the plant. I recommend a 4 or 5 gallon sprayer. For a very small vineyard, a 1 or 2 gallon hand-held spray pump will do.

Spraying is not that difficult, but it's important to be thorough. You can purchase different nozzles to change spray pattern, and change the angle to reach difficult spots. I like to use a nozzle that allows the spray to thoroughly cover the leaves and grape bunches. It's important to get most disease control agents into the cracks and crevices of the grape clusters and around the fruiting zone, and to get good coverage on the foliage.

A backpack sprayer makes the work of spraying the vineyard go more quickly. Be sure to wear gloves and goggles to protect yourself from the fungicide. FACING PAGE: Thoroughly cover all the vine surfaces with the spray, including the young grapes.

In some cases you will want to add a "no foam" or spreader/sticker product that helps the spray to stick to the leaves and fruit; many commercial formulas will already have such an agent included.

With backpack sprayers there is little overspray or residue. Be sure to triple rinse your spray rig with water after use, so it is ready to go and clean for the next application. I also recommend using separate spray rigs for herbicides and disease applications, just in case some minor herbicide residue remains.

THE LABEL IS THE LAW

The label on a product container is a legal document that describes exactly how the product is to be used, based on extensive testing. Make sure you understand how to mix and apply the product before you begin, then follow all label directions very carefully and exactly.

Also follow label directions for wearing protective gear when mixing and applying products. This may include plastic or rubber gloves, safety glasses or goggles, a respirator rated for products, long-sleeved shirt, long pants or coveralls, and closed shoes (no sandals or bare feet).

GRAPEVINE **DISEASES**

Anthracnose

Elsinoe ampelina

In rainy, hot, and humid areas, anthracnose can take hold and cause damage to vineyards. The fungus overwinters on infected shoots and then spreads by rain onto new tissues in spring. Hot weather encourages growth. It affects all parts of the growing vines, but you can recognize it mainly by the lesions that form on shoots and berries. These start out as small reddish circles that then grow and become sunken and gray. Darker margins form around the edges of the lesions, and the tissue may crack. Leaf spots are similarly circular with gray centers and darker margins.

Finally, the fruit develops the small circular spots, which enlarge and turn whitish grey with reddish to black edges. (These spots look like bird eyes, and the disease is sometimes called bird's eye rot.) Eventually the fruit cracks and splits.

CONTROL

Prune out and destroy infected wood. Clean away fallen foliage, fruit, and prunings. Plant resistant varieties. Manage the canopy to encourage air circulation. Apply a lime sulfur-based fungicide as a dormant spray in late winter, before bud break, then through the growing season at two-week intervals.

Black Measles (Esca)

Togninia spp.

Black measles symptoms show up mainly in midsummer to late summer. Leaves develop dry spots between veins. Leaves can drop off, and tendrils and canes may die back. The disease enters through pruning wounds. It is most prevalent in areas with hot summers and where spring rains are heavy.

CONTROL

Some experts suggest that control of black measles can be achieved with use of liquid lime-sulfur applied after pruning. However, you must get the product into the cracks and crevices of the vine, as well as into any cuts or pruning wounds, to attack the fungal fruiting bodies. Other treatments include use of wax or tree tar to fill pruning wounds or other holes. Though this method is still experimental, there would be no way for the fungus to reinfect the vine if these holes are plugged up.

Black Rot

Guidnardia bidwellii

Black rot is caused by a fungus that overwinters in mummified fruit on the vine or the ground. When fungal spores are released during spring rains, leaves may be reinfected. Brown circular lesions appear on the leaves with small black dots within the lesions. Shoots, cluster stems, and tendrils may also develop lesions. Fruit can be infected during blooming and eventually turns dark brown before shriveling into raisinlike mummies.

CONTROL

Timing of fungicide is critical: from just before bloom to up to a month after bloom. However, copper and sulfur are not very effective against black rot, so good vineyard sanitation is called for. Remove mummified fruit, especially any that is hanging on the vine, and clear up infected leaves and canes. Do not compost the material.

Bunch rot, or botrytis, can appear late in the season and attack grapes just as they are ripening.

Bunch Rot

Botrytis cinerea

This fungal disease, also called botrytis or gray mold, can grow on any plant material, but mostly on young shoots or flower parts, on stressed or ripened fruit, or on dead (yellow) leaves. This disease damages ripened fruit and can account for severe losses. Zinfandel seems to be more susceptible to this disease than other varieties, because its very tight bunches prevent the interior grapes in the bunch from drying out.

The fungus overwinters on leaves and fallen fruit on the vineyard floor. Under moist conditions in the spring, the spores of this fungus can infect the grape flowers, young shoots, and young leaves. Symptoms of bunch rot are evident when the cluster turns brown or the skin slips off the berry. The characteristic sign of bunch rot is a fluffy, gray-brown growth on the fruit. Since these symptoms appear late in the season, treatment and prevention early on is important. The fungus can also infect berries through wounds, such as from birds.

CONTROL

You can best control bunch rot through an integrated approach that includes spraying, mechanical means (including canopy management), and sanitation and irrigation techniques.

Practice good canopy management to increase exposure of the grape clusters to air and sunlight, just as with powdery mildew. From early summer to late summer, pull leaves and shoots to expose 70 to 80 percent of the fruit. Avoid overhead irrigation, as moisture on the vines creates an

environment that allows bunch rot to occur. Encourage air circulation and keep the vines dry.

A regular program of preventive spraying with fungicides such as organic Stylet-Oil can help control bunch rot. While several methods of spray control have been suggested, studies show that spraying when environmental conditions are conducive to the growth of the fungus (during times of wetness or higher humidity in the vineyard) seem most successful. Just as with powdery mildew, begin your spray after bud burst and continue every two weeks through the damp or rainy part of the season, until veraison. Do not spray after veraison. If there was an infection in the previous year, do not leave infected clusters on the vine or on the vineyard floor.

Crown Gall

Agrobacterium tumefaciens

Vineyards in colder regions are more susceptible to this disease, which is caused by a bacterium. Both pruning wounds and damage caused by freezing can be sites for the bacteria to enter. The growths, or galls, usually form on the base of the trunk near the soil line. They eventually become dark, bulbous, and hard. They interfere with movement of water and nutrients to the rest of the vine, resulting in poor growth and eventually death of the vine.

CONTROL

Always inspect new vines for galls. Prevent freezing damage to vines by protecting them from winter injury. Control other fungal diseases. Remove diseased vines and destroy them. There is no recommended organic treatment for crown gall.

Downy Mildew

Plasmopara viticola

This fungal disease is a major problem for growers in the eastern U.S. Vinifera grapes are more susceptible to downy mildew than American varieties, and French hybrids are somewhat susceptible.

As with other fungal diseases, the spores for downy mildew overwinter on the vineyard floor. Humidity and splashing rain encourage the release of the spores, which infect green tissues on the vines. The affected leaves turn brown, spotted, or mottled on top. On the underside of the leaves is where you find the white, fluffy growth that is recognizable as downy mildew. Vines may drop infected leaves; shoots and tendrils can curl and thicken. Eventually, green tissue can die completely.

Fruit infected with downy mildew turns brown, then purple, and is covered with the white fluffy growths. The grapes detach easily from the stems.

CONTROL

Good prevention is important to help deal with downy mildew. Plant resistant varieties. Prune and train vines to encourage plenty of air circulation. Clean up debris between the rows and control weeds.

As with other fungal diseases, treatment for downy mildew should start before bloom and continue for a month after bloom, at weekly or biweekly intervals. Copper is the primary fungicide for treating downy mildew, including copper sulfate, Bordeaux mixture, and commercial organic preparations containing copper.

Eutypa Dieback

Eutypa lata

In spring, growing shoots show signs of the disease. Leaves may be small, stunted, and yellow. You may notice these symptoms for a few years until the vine actually starts to die back. But eutypa usually starts around old pruning wounds, forming a canker on the trunk. If you pull away the bark over the canker, you'll see a flattened, dark area up to 3 feet long. By the time you see this kind of damage to the trunk, the disease has progressed.

CONTROL

If you notice shoots that show eutypa damage, identify and remove infected vines. All wood should be removed and destroyed, as the fungus can continue to emerge from trunks. Prune late in the dormant season so that pruning wounds heal quickly. There is no approved organic treatment for eutypa dieback.

Phomopsis Cane and Leaf Spot

Phomopsis viticola

Cool weather and rainfall release overwintering phomopsis fungal spores from infected wood. Black spots and lesions appear on the shoots. Although the fruit is infected when young, it does not start to rot until later in the season. Close to harvest, the grapes might turn a light brown color, spotted with black.

CONTROL

Copper fungicides are recommended for phomopsis. The critical period is when the clusters are developing until up to a month after bloom.

Powdery mildew appears early in the season and can cause a white fuzz on young grapes and black spots and streaks on the canes.

Pierce's Disease

Xylella fastidiosa

Pierce's disease is a deadly bacterial disease of grapevines. The bacteria is spread by insects, most notoriously the glassy-winged sharpshooter. Once in the vine, the bacteria plugs up the xylem, which carries water through the plant. Leaves may have yellow or red margins, and eventually the leaf margins dry out in concentric zones. Fruit will shrivel, and the wood on new canes develops irregularly. In subsequent years, leaves become chlorotic and scorched. Shoots and cane tips die. The disease is more prevalent in hot areas and less so in areas with cold winters. Some varieties are more susceptible than others. Chenin Blanc, Sylvaner, Ruby Cabernet, and White Riesling seem to be the most resistant.

CONTROL

There is no treatment for Pierce's disease. If the disease is confirmed, remove and replant vines to maintain production. Plant resistant varieties on resistant rootstocks. Control insect vectors, especially sharpshooters.

Powdery Mildew

Uncinula necator

This is the most serious and widespread fungal disease in many grape-growing regions of the world, especially humid areas. Despite treatment, powdery mildew in some years can result in losses. The degree of mildew varies between different varieties, with vinifera and French hybrids being more susceptible. But even within this group, certain varieties are more prone to powdery mildew. For instance, Carignan, Chardonnay, Cabernet Sauvignon, and Chenin Blanc are more seriously affected, while Petite Sirah, Zinfandel, and White Riesling are less susceptible.

Powdery mildew can affect all the tissues of a grapevine. It appears as a white powdery covering of leaves and, when severe, on the grapes. Leaves may curl, especially in hot weather. Late in the season small, spherical, black fruiting bodies may form on the leaves, shoots, canes, and grapes. Severe mildew can cause berries to crack and split, allowing rot organisms to enter. Powdery mildew cannot grow on dormant grape tissue but can survive the winter under infected buds.

CONTROL

Management of powdery mildew is done mainly by prevention. Practice good canopy management techniques such as keeping an open canopy. This allows the sun and wind to dry the vines to prevent dampness. I recommend leaf pulling and shoot thinning from early summer to late summer.

Sulfur has historically been the choice of sprays for powdery mildew. I use Stylet-Oil and in the early part of the season add copper to the

application. Begin your powdery mildew spray after bud burst and continue every two weeks through the damp or rainy part of the season, until veraison. When you have grapes to harvest, do not spray after veraison.

Sour Rot

This disease is caused by a mix of yeast and bacteria and is often found with other fungal diseases. The condition affects some varieties more than others: Pinot Blanc, Pinot Gris, Pinto Noir, Riesling, and Chardonnay are all susceptible.

Fruit discolors and rots, with leaking juice and a smell like vinegar. You will usually find fruit flies and larvae in large numbers. The condition seems to enter grapes through wounds caused by other diseases, or by wasps, birds, or other sources.

CONTROL

Avoid making large pruning wounds during early winter rains. In milder climates, you can prune late in winter, which tends to prevent most wintering pests and diseases. Organic barrier products applied to pruning wounds may help prevent infection during hazardous periods. Bordeaux mixture applied around veraison may help. Control wasps and birds. Remove affected grapes and destroy.

Deer are one of the biggest threats to vineyards even in suburban backyards. You need a sturdy, tall fence to keep them at bay. And don't forget to close the gate at night.

VINEYARD PESTS

A number of common animal pests cause vineyard damage in almost every region of North America. Deer, for instance, are a widespread and growing threat, even in suburban or urban neighborhoods. Likewise, most growers find they have to deal with hungry birds. In the West and Midwest, gophers may be the biggest problem, whereas in the Northeast, raccoons may cause the most damage. You will soon learn which animals are preying on your vineyard and how to deal with them.

Deer

Deer can devastate a vineyard. If your vineyard is vulnerable to grazing deer, you must either build a fence or have your vineyard devoured by them. Deer will most often come into populated areas when the green grazing plants in their previous neighborhood are depleted. That is when they will eye your vineyard. Deer mostly feed at night, but when they are hungry they can be quite brazen in your vineyard at any time of day. They can strip your vines down to the canes, and are especially fond of tender young shoots.

If you have horses, cattle, llamas, or other large grazing animals, they can cause the same damage as deer—and they will make a mess of the vineyard floor as well.

The best prevention is a tall fence. I live in a semi-rural area and a 5- to 9-foot fence keeps out the deer. In more rural areas, I recommend at least a 7-foot fence. A traditional wire deer fence with gates that close firmly is the gold standard. However, depending on the layout of your property, that can be an expensive proposition.

Newer materials that work for many homeowners include fences of mesh or plastic netting. Various electric fencing options are also available; some have single-strands that deliver a mild shock that trains the deer to stay away. Even fishing line or steel wire stretched across a small area can sometimes deter local deer. Make sure that the wire is under tension, so that it does not give way when the deer inspect it.

In a small suburban yard where there are other options for the deer, sprays may provide some protection. There are a number of sprays and recipes on the market, most including some combination of eggs, predator urine, garlic, or meat. Sprays containing sulfur compounds are said to resemble an alarm scent to the deer. The sprays will need to be applied consistently and frequently to have any effect.

A barking dog will also help, but do not depend on just the dog as your only form of protection. If the dog leaves for a while each day or falls asleep for a while, it only takes a short time for deer to completely de-leaf your vineyard.

Rabbits and Squirrels

Small mammals can also eat small or newly planted vines, so watch out for them. Once your vines have begun to form woody main trunks, the pests will tend to leave them alone, preferring the tender young shoots. Grow tubes will protect your developing vines during the first year. Repellents are also available. Small vineyards may be enclosed with rabbit-proof fencing that extends into an underground trench at least 6 inches deep.

Raccoon, Rats, and Opossums

These animals pose little danger to shoots and leaves, but they will go for the fruit, especially when it is ripe. Foxes and coyotes are also said to poach grapes in some areas. Your best means of control for these animals is a fence—although raccoons are good climbers. Traps may be your only recourse for habitual offenders.

Gophers and Voles

Gophers are very damaging critters in a vineyard. They move over ground at night and often arrive in the vineyard when there are large, lush clumps of weeds with big roots. The water from the drip system often attracts gophers during dry periods. They dig burrows under the vines and can eat away young vines from the roots. Sometimes you do not even know it until the vine literally collapses. In addition, they will sometime gnaw on underground irrigation pipes, causing expensive and frustrating damage.

Keeping down weeds can help to prevent these burrowing pests, but if you live in an area with gophers the most you can hope for is control, as I do not believe you will ever eliminate gophers. I use standard gopher traps as the main methods of control.

Owl or other raptor perches or boxes are also good additions to the vineyard, as many feed at night and get the gophers while they are out of their holes. Perches and boxes can be found at agricultural stores or in organic garden books or stores. I have six owl boxes positioned around my vineyard, built specifically for barn owls. A barn owl family can eat several dozen gophers and other rodents per night.

Birds

Birds are beautiful in the vineyard (and can be beneficial for the rest of your garden) but not when you have a crop of ripe grapes. Birds eat grapes, and the more birds you have the more grapes you will lose. During the period before ripeness, they will not spend a long time in the vineyard, but as the grapes ripen, you will see them stay around longer. I find starlings the worst, but there are many birds that will feast on your vineyard. Even wild turkeys can enter the vineyard and peck at young shoots

Gophers can be a huge threat to your vines and even to your irrigation lines. Constant vigilance is needed to keep them down to manageable numbers. Owls and other raptors can help. Raccoons are great climbers and will put in considerable effort to reach your ripe grapes.

and, especially, fruit. Evidence of their interest in your ripe grapes will be broken berries and half-eaten bunches. Vineyards located near large stands of trees tend to attract more birds than areas with few trees. Bird populations vary from year to year; one year you may not have a serious problem and the following year your crop could be under serious threat.

The least expensive way to discourage birds is to tie strips of reflecting tape in 4- to 5-foot lengths throughout the vineyard. This tape, which comes in small rolls, is often silver and red or other bright colors, and can be purchased at a nursery or garden center. Tie the tape in random locations throughout the vineyard, just before the grapes begin to ripen. Putting the tape out too early will allow the birds to become accustomed to the tape and it will not be as effective.

You can also try pre-programmed recordings of predatory bird calls, which differ in variety. These can be purchased in vineyard supply stores. Other scare devices are sold at farm supply stores and vineyard suppliers. They may include propane noise cannons, large inflatable balls, or replicas of predators. Although these methods may work, you will need to move them around and vary your tactics as the birds get used to them.

If birds are an ongoing problem in your area, you may have to net your vines. Netting is a pretty expensive proposition for large vineyards, but for small vineyards it is fairly affordable and it's a sure thing. There are several options for netting. The least expensive is extruded plastic netting. It comes in panels that are either 12 to 16 inches wide or 4 to 16 feet wide, and in lengths of 100 and 200 feet. The usual color is black. You can find this kind of netting, which may last up to three years, at nurseries or vineyard suppliers. To use it, place the narrower panels on either side of the fruiting zone and clip them together at the top and bottom. (You can also use wooden skewers, just thread them through the edges of the netting.) The wide panels can be draped or fastened over the entire row.

A more expensive netting product is a green or white nylon mesh. It will outlast the extruded plastic, usually lasting for seven or eight years. This type of netting is sold in rolls of 500 or 1000 feet, in widths as narrow as 4 feet or as wide as 20 feet. The narrow panels are best used in the same way as the plastic netting, on either side of the fruiting zone. Wider netting, say 12 to 14 feet, can be draped over the entire row of vines and clipped at the bottom. The widest 17-foot-and-up netting can be draped over the row and securing to the ground with U-shaped pegs. Applications of long rows of netting can be done with a dispenser on the back of a tractor, if you can borrow one, or can be applied by two people, one working down each side of the row.

Where birds are abundant, be sure the net does not touch the grapes as some birds use the net as a landing platform or ladder to eat your grapes. I prefer the widest kind of netting for this reason; it keeps the birds far away from the fruit.

There are several different ways to net your vines to protect them from hungry birds. You can drape netting over the entire row (**TOP**), or fasten it over the upper part of the vine (**BOTTOM**).

Owl Box

You can construct your own owl box from a plan that is specific to the types of owls most likely to be found in your area.

Securing Bird Netting

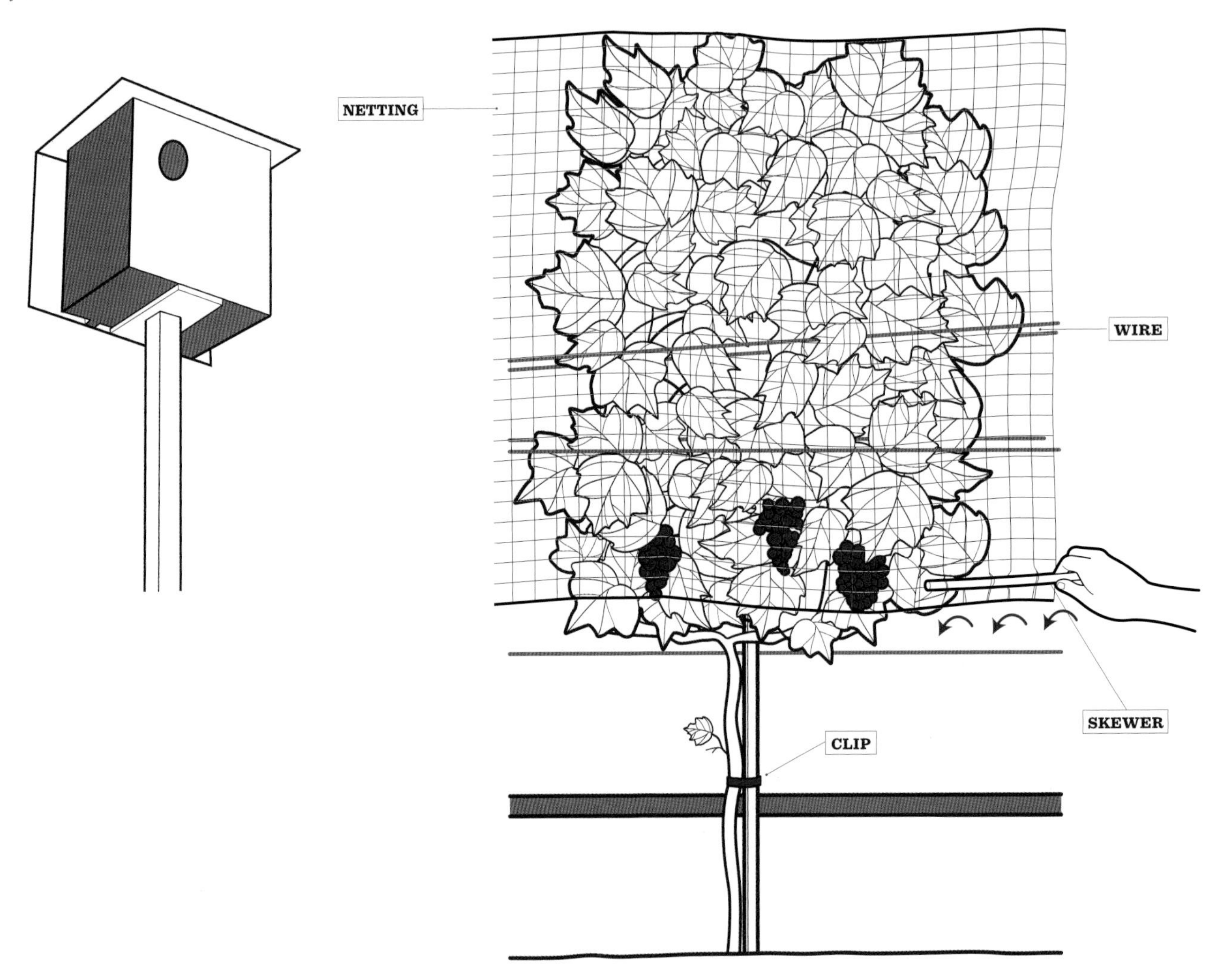

INSECTS IN THE VINEYARD

Most insects found in organic vineyards are neither very destructive nor even destructive at all. However, at some times in some places, your vineyard can be subject to insect damage. This varies greatly from one area to another, so local sources of information are especially helpful when dealing with insects. Check with other local growers, your extension service, or local agriculture office for more information on insect pests specific to your location.

Don't let insect pests catch you by surprise. Observe the vineyard on a regular basis so that you can spot damage early. If possible, catch samples of suspect pests; sticky traps are a useful tool. Put them at several points around the vineyard for the best sampling. As always, make sure you have correctly identified an insect pest before you treat it. Fortunately, there are excellent online sources for pest identification. Among the best are UC Davis and Cornell University.

Beneficial Insects

One of the best ways to control insect pests is by using natural predators. Get to know these insects so that when you see them you are not tempted to squish or spray them. They will help to keep your vineyard healthy. Cover crops, ground covers, and surrounding native habitats will encourage these and other beneficials that are indigenous to your area. As a bonus, you will also encourage bees that are helpful in pollinating the grapes.

I have planted a hedgerow along the side of my vineyard. It contains native plants that flower all times of the year, and they attract beneficial insects like bees, hornets, and spiders. I have noticed the local bee population thrives near my vineyard as a result of this hedgerow.

Domestic fowl are another good source of insect control. Chickens and ducks will all happily wander through the vineyard picking at insects in the ground. Geese are primarily vegetarian, but they will often pick up slugs and snails, and they may help to deal with weeds as well.

Organic Insecticides

Many gardeners and homeowners are in the habit of spraying every pest they see with a chemical spray. As I have said before, I do not recommend this approach, and especially not in the vineyard. First, most vineyard insects do not cause enough damage to affect your crop. Second, I believe that you can control most insects through a combination of beneficial insects, good vineyard sanitation, and patience. Still, there are times that you may be obliged to treat an invasion of pests with an insecticide.

Traps are among the means you can use to control insects. They usually contain an attractant that draws the insects. In fact, such traps can attract more insects to the vineyard than you would like.

If you do feel compelled to use an insecticide, remember the process I described: observation, identification, diagnosis, and then treatment. Make sure you have the correct product for the correct pest and that you are applying it when it will do the most good—whether at the adult or nymph stage.

Organic standards vary in terms of what is considered an acceptable insecticide and these standards may change. Listed here are some of those most commonly approved, but be aware that not all organic certification bodies consider all of these to be acceptable. Note that commercial formulas may combine different insecticidal substances with a synergist (a chemical that enhances an insecticide's effectiveness) such as piperonyl butoxide.

COMMON ORGANIC INSECTICIDES

Ammonium carbonate
Aqueous potassium silicate
Bacillus spp. (bacteria)
Bordeaux mixture
Copper sulfate
Diatomaceous dust
Elemental sulfur
Horticultural oils
Insecticidal soaps
Kaolin (Surround or other proprietary spray)
Lime sulfur
Neem
Nicotine sulfate
Pyrethrins
Pyrethrum
Rotenone
Sabadilla
Spinosad
Sticky traps

There are also traps to manage insects. Pheromone-based traps emit hormones that attract insects during the mating cycle. These traps contain an attractant meant to lure the insects into a trap from which they cannot escape. Sticky traps and tape work in much the same way and can catch a wide variety of insects. They can be helpful to identify other insects that may be pests. Note, though, that such traps can

COMMON INSECT PESTS

Here are some of the bugs that may eat your grape leaves—and even the roots. Only treat them with insecticides if the infestation gets too large to handle by other means.

Branch and Twig Borer

Also known as the grape cane borer, this pest is a dark brown beetle roughly cylindrical in shape. The larvae drill into wood on trunks and canes, often at the site of old pruning sounds. Shoots wilt and collapse.

CONTROL

Remove dead wood at pruning time. Burn brush or wood piles before larvae hatch in March. Beneficial nematodes may help control larvae; contact your local extension service for advice.

Flea Beetles

Native to eastern North America. Was once highly prevalent in New York vineyards. Small metallic blue insects appear early in spring. The beetles jump when disturbed. In summer, larvae and adults feed on leaves, but overwintering adults do the most damage in late spring and early summer, as they attack developing buds.

CONTROL

Yellow sticky traps placed along the row may capture emerging beetles in spring. You can try to hand pick beetles or knock them into a bucket of soapy water, but they are tiny and fast. Treat summer adults with pyrethrum or diatomaceous earth. Control weeds.

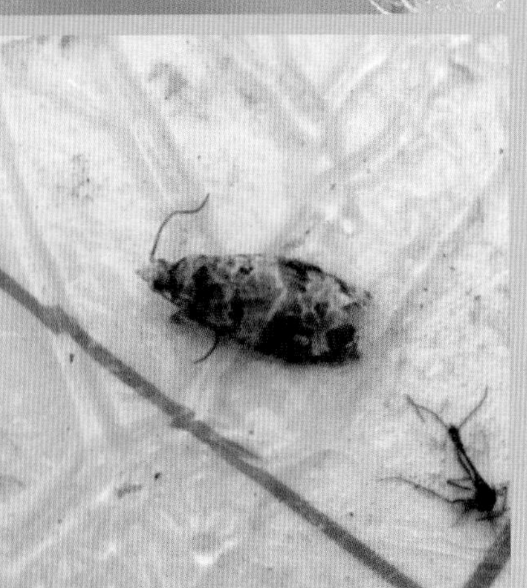

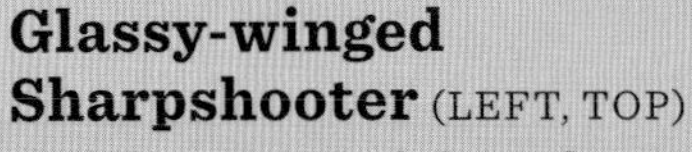

Glassy-winged Sharpshooter (LEFT, TOP)

The glassy-winged sharpshooter is almost 1/2 in. long and is dark brown with a lighter underside. In 2001 it became a serious new pest in California. Native to the southeast United States, this leafhopper was first seen in California in 1990 and is now found throughout much of the state. It is particularly a threat to California vineyards due to its ability to spread *Xylella fastidoisa,* the bacterium that causes Pierce's disease.

CONTROL

Encourage beneficial insects. If sharpshooters are a persistent problem, use a proprietary spray containing kaolin, a naturally occurring clay (Surround is 95 percent kaolin). There is no organic insecticide that will kill this pest; you can only try to block the insects from feeding on vines.

Grape Berry Moth (LEFT, BOTTOM)

Native to eastern North America; typically found east of the Rockies. Small, brown adults are about 1/4 in. long with a 1/2 in. wingspan. The moths lay their eggs in spring on buds and young berries. the larvae emerge and feed on shoots, stems, leaves, and young berries, often entering the berries and destroying them or making them susceptible to rot. Look for small red spots on the berries where the larvae have entered. Moths emerge in May and June, then a second generation in late summer through fall. First generation may pupate on the vines; in winter, the second generation pupates on fallen leaves.

CONTROL

Hand pick damaged berries. Clear fallen leaves from the vineyard floor in fall. Pheromone traps may disrupt mating.

Grape Cane Gallmaker

Small, reddish brown weevils can damage new shoots in spring, close to bloom time. Found in eastern and midwestern vineyards. The female lays her eggs in canes in spring. The larva eats the shoot, forming a 1 to 2 in. long gall as it pupates. Canes may break at the point of the gall. The grape cane girdler is similar.

CONTROL

Prune out visible galls. Clear fallen leaves and debris in late fall.

Grasshoppers

Some species become active in spring, eating young shoots; others are destructive in late summer when surrounding grasslands have dried up.

CONTROL

Encourage natural predators, including praying mantis and insectivorous birds, although you may need to net your vines to protect the ripening grapes from the birds later in the season. Till soil to expose eggs to predators. Plant cover crops. Treat with diatomaceous earth, neem oil, or pyrethrins.

Japanese Beetles (RIGHT)

Familiar to gardeners in eastern North America, Japanese beetles are metallic-looking green beetles that can swarm over plants in early summer, sometimes decimating foliage. June beetles are similar, but larger and reddish brown.

CONTROL

Beneficial nematodes may help control larvae and reduce populations. *Bacillus thuringiensis* (Bt) and milky spore (*Bacillus popillae*) can treat larvae. Check with your local extension service for new beneficial insects to control adults. Pheromone traps tend to encourage more beetles, so position them away from your grapes. Pyrethrins and diatomaceous earth may help kill adults.

Leafhoppers

All leafhoppers are shaped like elongated triangles, with large heads. Several species feed on grapes: the grape leafhopper, the three-banded leafhopper, the potato leafhopper, the variegated leafhopper, and sharpshooters.

CONTROL

Parasitic wasps from the genus *Anagrus* prey on some leafhoppers, as do lacewings, minute pirate bugs, lady beetles, spiders, and mites. Encourage these beneficials. Control weeds to reduce the numbers of leafhoppers. Check your vineyard for leafhoppers and if their numbers become very high, treat with pyrethrins.

Nematodes

Although there are beneficial nematodes that are used to treat some soil-dwelling grubs, there are also harmful nematodes that can affect the vineyard. These microscopic roundworms can feed on plant roots, puncturing and sucking out the inside of individual cells. Several different kinds affect vineyards, including root knot, dagger, citrus, lesion, and ring nematodes. Some types are more prevalent in certain soils.

Nematode damage can resemble that caused by other conditions. Vines may lack vigor and have restricted growth or poor yields. There may be galls or swellings on the roots. To determine whether you are dealing with harmful nematodes, you will need to take soil samples and send them to a commercial laboratory for identification.

CONTROL

No single rootstock is resistant to all nematodes, but several are resistant to one or more types. Use manures and compost to improve soil health and vine vigor. Healthy, well-irrigated vines are less susceptible to the effects of nematodes. There are several chemical treatments for nematodes, but none that are acceptable for an organic vineyard.

Orange Tortrix (RIGHT, TOP)

A bell-shaped, ½ in. flying insect most often found in coastal vineyards on the West Coast. It feeds on shoots, leaves, and even grapes that have been left on the vine from the previous year. The larvae burrow into growing fruit and make nests of webbing in the cluster. A similar moth, garden tortrix, may be caught in traps but is not a grape pest. The chevron pattern on orange tortrix has less contrast than that of the garden tortrix.

CONTROL

Close monitoring is recommended for this potentially serious larval pest. Inspect weeds and vine shoots early in season after flower clusters become attractive to larvae. Encourage beneficial insects, especially parasitic wasps and spiders. Clear weeds and remove mummified grapes from vines. Pheromone traps may be effective in midwinter. Treat with *Bacillus thuringiensis* or spinosad.

Phylloxera (RIGHT, BOTTOM)

This infamous pest of grapevines has caused havoc in vineyards for many years. When it was unintentionally imported into France in the 18th century, it wiped out entire vineyards. Phylloxera is a louse, a small winged or wingless, oval insect that may be bright yellow, orange, brown, or green, depending on the stage of its life cycle. Nymphs feed on leaves, forming galls. Root-feeding phylloxera also forms galls, eventually killing the vine.

CONTROL

Use resistant rootstocks for control. Do not plant in soil that has previously been infected with phylloxera. Limit plant stress by building healthy soil and giving vines adequate irrigation and fertilization.

Thrips

These tiny insects have distinctive feathery wings. They feed on emerging shoots, leaves, and fruits. Stunted shoots with bronzing in early season indicate a damaging population of thrips.

CONTROL

Encourage beneficial insects such as minute pirate bugs. Treat with Stylet-Oil or spinosad.

Spider Mites

Mites are tiny green, yellowish, or reddish pests that form characteristic silken webbing on leaves. Damage may start as yellow spots that lead to bronzed and burned-looking leaves.

CONTROL

There is no prescribed treatment time, but populations are usually high in later summer. Encourage beneficial insects such as predatory mites. Reduce dust in the vineyard. Treat with insecticidal soap, neem oil, horticultural oil, or a product labeled for use on mites such as Stylet-Oil.

Vine Mealybug (LEFT)

Small, 1/8 in. long, flat, oval insects covered with a white, mealy, waxlike substance. Eggs, crawlers, nymphs, and adults may all be found under bark, on buds, and around the vine roots. Plants may have large amounts of honeydew on the leaves that results in black sooty mold.

CONTROL

Make sure to identify the mealybug (there are types other than the vine mealybug) and, if present in large numbers, release the lady beetle known as the mealybug destroyer to treat eggs and crawlers. Oil- or soap-based insecticides may control adults.

Weevils (RIGHT)

Weevils are small, 3/8 in. long, dark gray or brown insects with distinctive snouts. There are several different kinds that affect vineyards. The larvae feed on grape roots through most of the growing season. Adults may chew leaves, but the larvae do the most damage.

CONTROL

Beneficial nematodes may help control larvae; apply late in the growing season. Use compost to build up soil quality. Treat adults with a proprietary spray containing kaolin, a naturally occurring clay (Surround is 95 percent kaolin), diatomaceous earth, or pyrethrins.

attract more insects to the vineyard than you would like. Place them on the perimeter to draw the insects away from your vines.

Again, I would encourage you to try all means of managing insect pests before applying any insecticide, and to make sure that any product is labeled for the pest you are trying to treat.

Chickens can do double duty in the vineyard. They eat large numbers of insects, and they provide manure as they peck their way through the rows.

BENEFICIAL INSECTS

BENEFICIAL	APPEARANCE	TARGET PESTS
Assassin bugs	Slender, often colorful insects that may be red, black, brown, or greenish, depending on species. Long bodied, with splayed legs and a long narrow head. Nymphs are $^1/_4$ in. in length.	Many small- to midsized insects, including aphids, caterpillars, and leafhoppers.
Damsel bugs	Gray or brown, long, slender insects about $^2/_5$ in. long. Fast-moving.	Many small insects, including caterpillars, leafhoppers, mites, and thrips.
Lacewings	Adults are green or brown, with tear-shaped membranous wings. From $^1/_4$ to $^1/_2$ in. long, depending on species. Larvae look like miniature alligators with reddish stripes. Silken cocoons are attached to leaves or bark.	Many small insects, including mites, leafhoppers, and mealybugs; also eat insect eggs.
Lady beetles	Small, $^1/_{16}$ in. long familiar garden beetles may be brown, black, or spotted. Larvae are white or black, like miniature alligators with long legs.	Both adult and larvae will feed on different bugs, depending on species. Pest-specific lady beetles include mealybug destroyer and spider mite destroyer.
Mantids (praying mantis)	Distinctive, 2 to 4 in. long yellowish, green, or brown insects.	Eat many pest insects, but also other beneficials. Introduce with care.

BENEFICIAL INSECTS *continued*

BENEFICIAL	APPEARANCE	TARGET PESTS
Minute pirate bugs	Small (less than 1/5 in. long), dark purplish oval insect with white markings and triangle-shaped head. Nymphs are yellowish or reddish-brown with red eyes.	Both adults and nymphs prey on many small insect pests and eggs such as thrips and mites.
Nematodes	Microscopic soil-dwelling roundworms.	Control insects in larvae stage, including weevils, beetles, and caterpillars. There are also nematodes that can be damaging to grapevines.
Parasitic wasps	Typically smaller than familiar garden wasps, brown or black in color.	Many types are effective against various pests. If you purchase wasps from a commercial source, be sure that you have the correct species for your particular pest problem.
Predatory mites	Tiny, cream to reddish, pear-shaped insects.	These eat other mites. The western predatory mite feeds on spider mites.
Spiders	Many types, with 8 legs and 2 main body parts	Eat a wide variety of insects. Most are harmless to humans, but several can cause severe bites.

Lacewing

Praying mantis

Lacewings are among the many beneficial insects that you want to encourage in your vineyard. These delicate little flyers can consume not only mites, leafhoppers, and mealybugs, but they'll even hunt down and consume insect eggs as well.

Praying mantids are another voracious eater of larger pests, but they may also prey on beneficial insects, so release them with care.

WHEN SHOULD I **HARVEST?**

BALANCING **SUGAR** -AND- **ACID**

HARVEST Time

HARVESTING YOUR GRAPES

STORAGE and WINEMAKING PREP

When the grapes are ripe and ready to harvest, start calling your friends to recruit a team of pickers. These Dornfelder grapes are ripening in a Finger Lakes, New York, vineyard.

WHEN SHOULD I HARVEST?

This is a good question. The harvest season can vary widely depending on your location, the weather, and the type of grapes you are growing. In the warmest regions, harvest can start in August; in cooler areas, it may commonly take place in October. Late harvest wines are usually picked in November, and icewine harvest can happen in December or even January.

Unlike some other fruit, grapes do not continue to ripen after they have been harvested, so they must be picked at just the right time. During veraison, the sugar levels in the fruit rise thanks to the stored carbohydrates in the plant and ongoing photosynthesis. At the same time, the acid levels in the fruit drop. The balance between sugars and acid determines the best time to harvest your grapes. If you are an experienced winemaker, you will have a pretty good idea what you are looking for in sugar content, pH, acid balance, and a few other things. If you are a novice, I can give you a few basic concepts that growers follow to comply with what winemakers want.

BALANCING SUGAR AND ACID

You will take two measurements to determine when to pick your grapes.

1. The sugar content, measured by taste (the old-fashioned way) or in terms of Brix, as measured by a refractometer or hydrometer: 1 degree Brix is 1 gram of sugar in 100 grams of solution (grape juice). The range for grape sugar is:
   - From 20 to 25 Brix for regular table wines
   - From 26 to 33 or more Brix for late harvest or sweet wines (port, sherry, and the like)

2. The pH levels, measured with a pH meter:
   - A range of 3.1 to 3.3 for white wines
   - A range of 3.4 to 3.5 for red wines

   Grapes below 3.1 will be too acidic. Above 3.5, the wine will be subject to oxidation and bacteria.

You may also want to measure acid levels, which we'll consider next.

Wine Acidity

Although most of your work in the vineyard has been to manage the sugar levels in the grapes, it's also important for growers and winemakers to understand and manage acid levels in the grapes.

There are several types of acid in wine, including primarily tartaric acid, malic acid, and citric acid. The levels of acid in the grapes should be different depending on the style of wine you plan to make, and for individual tastes. The measurement of acidity that that indicates the strength of the acid in the grapes is the pH. It is measured using a titration kit or pH meter. TA, or titratable acidity, represents acidity as a percentage of volume. You can use the pH meter to measure pH but not TA; for this you need to use a lab for testing, or buy a relatively expensive and complicated setup to test TA. Home winemakers may not measure this factor with such precision, but commercial wineries do, and you can have your grapes tested by a commercial lab if you choose. A pH meter, on the other hand, is inexpensive and easy to use.

Appropriate acid levels are:

- A range of 0.65 to 0.85 TA for white wines (dry white wine 0.65 to 0.85; sweet white wine 0.70 to 0.85)
- A range of 0.60 to 0.80 for red wines (dry red wine 0.60 to 0.70; sweet red wine 0.65 to 0.80)

There is a relationship between sugar, TA, and pH level. For instance, the perfect Cabernet Sauvignon balance of these three factors is 22 to 25 Brix, 0.75 TA, and a pH of 3.4. If all those numbers are in the correct range, the grapes will produce a good wine for the variety with little intervention. If the numbers are too far outside the ideal range, the only way to make adjustments is during the winemaking process, but you are less likely to produce a quality wine.

It's a good idea to develop a testing process from veraison to harvest that determines Brix and pH. Along with grape color and taste, these two measurements are usually sufficient. You can also consult with the winemaker if you are selling your grapes. At some point you may be able to accomplish what the grand winemakers do—simply taste the grape in the vineyard and know when to pick.

MEASURING BRIX

Refractometers are not expensive, but they do contain fairly sophisticated optical equipment. Essentially, the refractometer uses a prism assembly to read the amount of sugar in a liquid sample that is sandwiched between a daylight plate and the prism. The reading, in Brix, is indicated on a scale.

Calibrate your refractometer following the manufacturer's instructions (refractometers are calibrated to work at a specific temperature). Once the instrument has been calibrated, clean the daylight plate and the top of the main prism assembly with a soft, damp cloth.

Pick a cluster of grapes and squeeze the juice into a sterile bowl, then strain the juice. Place 2 or 3 drops of the juice onto the top of the prism. Close the daylight plate and let the juice spread out over the prism for about 30 seconds. Hold the instrument up to a natural source of daylight and look into it, focusing the eyepiece as needed. Read the scale and write down the Brix. Until you become very familiar with the instrument, be sure to take several readings. Always clean your refractometer with water after use and store it in a clean, dry place.

A hydrometer is a glass cylinder with a weighted bulb that measures the density of grape juice or must, which correlates to the amount of sugar in the liquid. You place the hydrometer in a beaker of must, and it will float at a certain level which is read according to a Brix scale on the floating cylinder. The hydrometer is a helpful tool to determine the Brix level, but it is not very portable and you need enough juice to fill up a beaker in order to get an accurate reading; thus it is more helpful for winemakers than for grape growers. A refractometer is a small hand-held device that can measure just a few drops of juice from the grapes, which means it can be used out in the vineyard and is more convenient for the small grower.

The refractometer is one of your most valuable tools. Use it to evaluate the sugar level of your grapes and decide when to harvest.

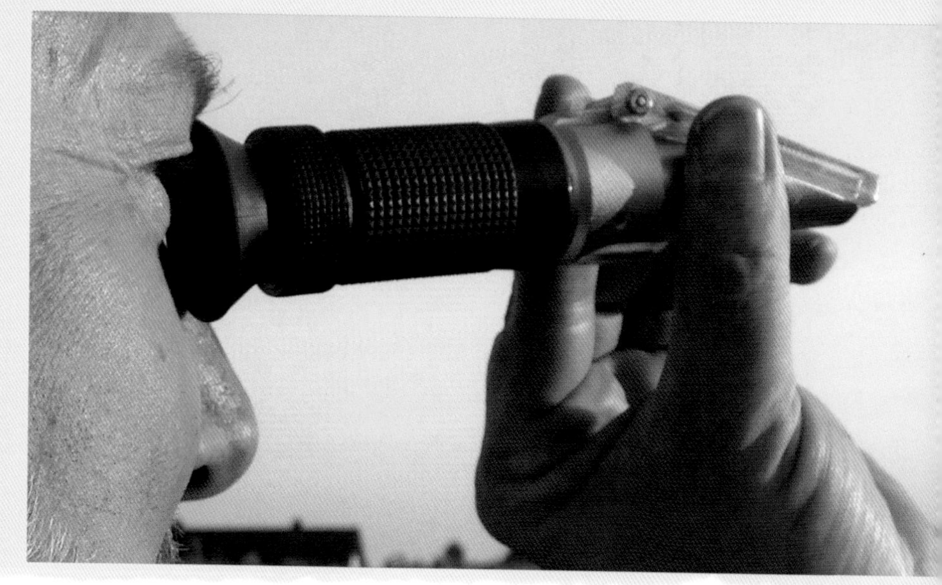

HARVESTING YOUR GRAPES

If you are going to sell your grapes or supply them to a winemaker, you may want to negotiate the cost of labor for picking the grapes as part of the deal. If you are like most small growers, though, you will want to make your own wine and perhaps have a harvesting party. This is the time that all your friends are waiting for. They want to see what you have done and what your vines have produced. Naturally you'll want to have some wine (from the year before), cheese, and some baguettes for your helpers.

Before harvesting, you may want to trim back shoots to facilitate harvest. But don't cut back too much; excessive pruning can delay ripening. Remember it is not sun on the grapes that ripens the fruit, but sun on the leaves.

HARVESTING EQUIPMENT

- **Picking knives**
  These are hook-shaped knives that you can get from a farm supply store. Use these knives carefully. They are sharp, and they can be dangerous.
- **Pruners**
  Regular garden pruners are safer and easier to use than a picking knife.
- **Picking bins**
  These may be 30 to 50 lb. wood, plastic, or metal bins, into which you drop the fruit after you have cut it from the vines. I like the 30 lb. plastic bin because it is light and sturdy. The bins are stackable and made of food-grade plastic, so they are easy to clean and store.

Picking the Grapes

Work your way along the rows. Use one hand to support each grape cluster and snip the stem with pruners in your other hand, leaving enough to make handling the grapes easier. The stems are quite tough, especially in some varieties like Petit Sirah, so be sure to keep your tools clean and sharp. If you are using a picking knife, the technique is to hold the cluster from underneath and quickly pull the knife through the stem toward you. Again, this is a sharp tool: be careful if you use it to harvest your grapes.

Place the cluster in the bin rather than throw it—you want to leave the grapes intact if possible. You may find it easier not to wear gloves, but be careful, especially if using a picking knife. Watch for wasps, which love ripe grapes, and protect yourself from the sun.

As the picking bins become full, transfer the grapes to your larger $1/2$ or $3/4$ ton macro bins where you will get them ready for transport to the winemaker, to the wine press, or for crushing/destemming. If you cannot do this immediately, or if your winemaking is on a smaller scale, keep the grapes in a cool, dry location, out of direct sun or, better yet, in a refrigerated room.

Post-Harvest Care

After harvest, you will let the vines gradually enter dormancy in the winter. You will ease up on watering; one generous watering four to six hours after harvest is usually sufficient, unless you have already had rainfall. Fall is also a good time to plant a cover crop.

Harvesting Equipment

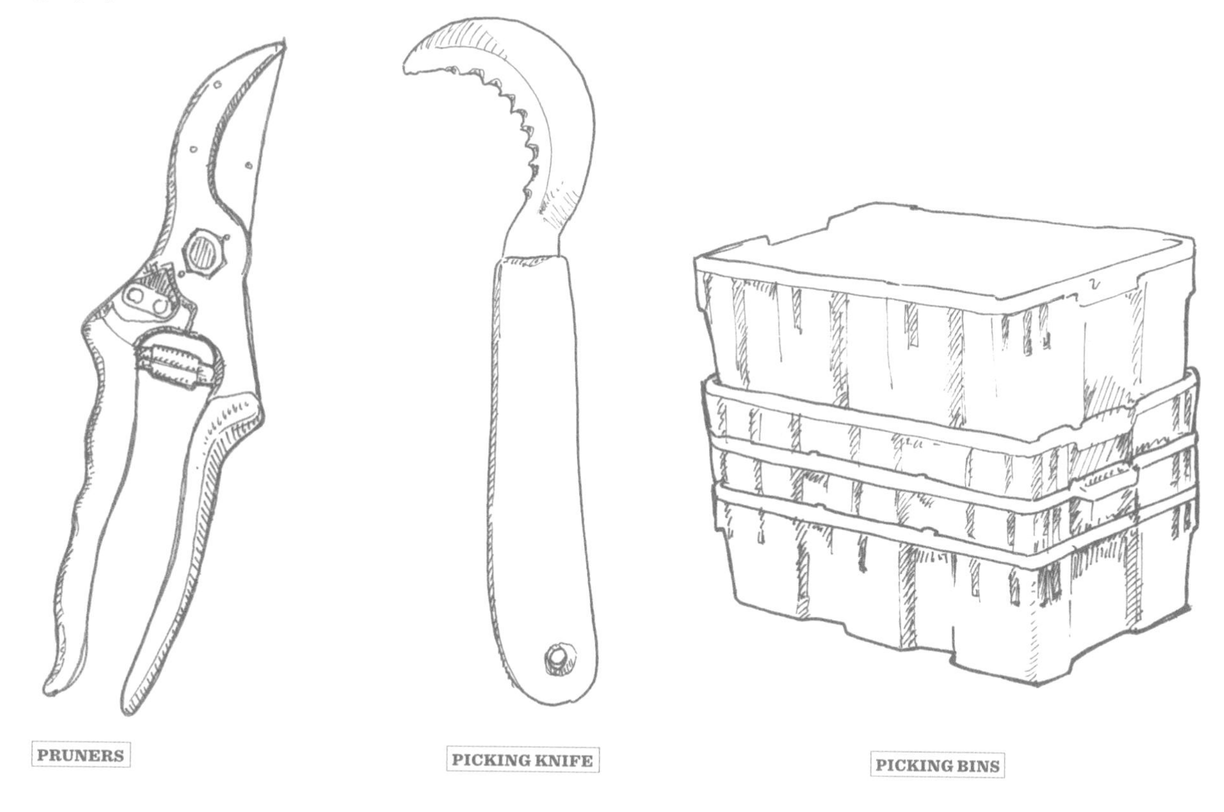

PRUNERS

PICKING KNIFE

PICKING BINS

Hold the clusters of grapes with one hand and snip off the stem with the pruners.

STORAGE AND WINEMAKING PREP

Whether you will be making your own wine or delivering your grapes to a winemaker, here are some tips to help your grapes make the transition from the vineyard to the winery in good shape. Wine goes bad mainly because of unsanitary conditions in the handling of the grapes, allowing the grapes to sit in a warm environment, or having grapes that are diseased (bunch rot is a frequent culprit). Here are some good practices for growers to maintain:

1. All picking and transport containers should be cleaned by a thorough washing. Use a solution of antibacterial soap or 5 percent hydrogen peroxide and water, scrubbing with a stiff brush, or steam clean with a power washer.

2. Pick grapes early in the day.

3. Pick only grapes that are free of disease.

LEFT: Plastic picking bins are easy to clean and to handle. **RIGHT**: Transfer the grapes to larger macro bins to prepare for processing.

Your reward for all the hard work in the vineyard: freshly pressed grapes in a barrel press. Be sure to put this pomace into the compost so that you return to the vineyard what you took out of it.

4. Keep the picking bins free of as many leaves as possible.

5. Store the grapes for a short time out of the sun in a cool, dry place or a refrigerated area. A few days' storage in a good, refrigerated environment can benefit the winemaker by preventing natural fermentation from beginning.

6. If you are selling your grapes, don't try to hide anything from the winemaker. If you have any concerns about the harvest, it is better to have it out in the open, when problems can still be corrected.

Glossary

ACID Measured as the pH level of the grapes, to determine when they are ready to pick. Total acidity by volume is titratable acidity (TA). This measurement is primarily used by professional winemakers.

AIR ROOT An outgrowth from the root system that emerges at the base of the trunk.

APEX The tip of a shoot.

ARM Horizontal growth trained from the top of the trunk, also known as the cordon.

AZIMUTH The arc of the sun's path as it crosses the sky.

B

BARK The dry, rough surface of a trunk or cordon.

BENCH-GRAFTED VINE The process of joining a scion to a rootstock in controlled conditions in a nursery or greenhouse.

BERRY GROWTH The period of time after flowering and through harvest when the fruit matures.

BERRY SET (ALSO KNOWN AS FRUIT SET) When berries begin to develop, usually 6 to 10 days after flowering. Some flowers fall off and those remaining develop into berries.

BLENDING Mixing together two or more different varieties to make a single wine.

BOTRYTIS CINEREA (ALSO KNOWN AS BUNCH ROT) A fungus disease that affects ripe grapes well into the season.

BRIX The percent of sugar in the grape, grape juice, or must.

BUD The initial swelling from a cane that will develop into leaves, tendrils, clusters, or other buds.

BUD BURST (ALSO KNOWN AS BUD BREAK) The point in spring where buds begin to emerge and unfurl from the canes. This usually occurs when the mean daily temperatures are 50°F (10°C).

BUNCH (OR CLUSTER) A group of grapes growing from a single stem.

BUNCH ROT See *Botrytis cinerea*.

CALCAREOUS CLAY A clay soil with high levels of calcium due to limestone or chalk. Believed by some to be the best soil for growing grapes.

CANE One-year-old wood that emerges as a shoot from the main trunk or cordons, and bears leaves and grapes.

CANOPY MANAGEMENT The process of physically manipulating the canes, shoots, and leaves to balance foliage and fruit growth and promote photosynthesis. Includes pruning, thinning and suckering shoots, positioning shoots between the upper training wires, and removing excess shoot growth around the fruiting zone.

CANOPY The upper portion of the canes consisting of the leaves.

CERTIFIED PLANT STOCK A plant that has been approved by a governing body to be healthy and virus-free.

CHLOROPHYLL The molecule in plant tissues that gives them their green color and absorbs light from the sun for photosynthesis.

CHLOROSIS Lack of chlorophyll in the leaves, which leads to fading and yellowing.

CLAY SOIL Soil characterized by small particles. Typically high in nutrients and water retentive.

CLIPS Stiff metal wires used to secure wires to the grape stakes.

CLONE An individual plant derived by vegetative propagation from a single original parent plant. Designated by a number after the varietal name.

COMPOST Decomposed organic matter that is used as a soil amendment to boost fertility and improve soil structure.

CORDON A permanent woody part of the vine that runs horizontally from the trunk and from which the fruiting canes emerge.

COVER CROP Low-growing annual plants grown between the rows to cover the soil.

CROP YIELD The amount of grapes produced by a vineyard, usually measured in tons per acre.

CRUSHING/DESTEMMING The stage in winemaking where the grapes are crushed to release juice.

D

DE-SUCKERING A process of cutting or pulling off unwanted shoots that emerge near the base of the vine.

DORMANT SEASON The period of time after leaf fall, when the photosynthetic and growth processes of the vine stop during the colder, darker months.

DRIP LINE The irrigation line, typically $1/2$ inch flexible PVC pipe, that supplies water to the vines via emitters.

DRIP WIRE The wire trellis system closest to the ground. The drip line is suspended from the drip wire.

E

EMITTER A plastic valve inserted into the drip line that emits a measured amount of water. In-line emitters are built into the interior of the drip line.

EVAPOTRANSPIRATION The process of water being drawn from the ground, through the vine's root system, up into the plant, and then out of the leaves.

EXCESSIVE VIGOR A characteristic of grapevines that have an excessive number of lateral shoots and too many large, thick leaves with long internodes.

F

FERTILIZING The process of supplying the nutrients needed for proper vine development.

FIELD GRAFTING The process of joining a scion to a rootstock while the rootstock is growing in the field.

FLOWERING When fruit bunch initiation or bloom occurs. Normally this takes place 30 to 80 days after bud burst, when average mean temperatures reach 61°F to 68°F (16°C to 20°C).

FLUSH There are two periods when roots put out significant growth, one in the spring and another in the fall.

FOLIAGE WALL A canopy that appears so dense as to resemble a wall.

FRENCH DRAIN A method to remove excess water in poorly drained soils. Typically, perforated 4 inch rigid PVC pipes are set in gravel at the bottom of a trench.

FRUIT SET Occurs about 10 days after flowering in midspring; the beginning of berry development.

FRUITING WIRE On the VSP trellis, the second wire up from the ground on which the cordon is trained and tied.

FRUITING ZONE The area above the cordon, where berries develop.

FUNGUS Organisms that cause common diseases of grapevines, including powdery mildew, bunch rot, and downy mildew.

G

GREEN-GROWING VINE A grafted grapevine delivered for planting after the leaves and roots are growing.

GROW TUBES Plastic or cardboard cylinders that are placed around newly planted vines to protect them from pest damage and to encourage growth.

GROWING SEASON The number of days between bud burst and leaf fall when the grape vine is actively growing.

GROWTH CYCLE The annual growth of a grapevine.

GUIDE WIRES The double set of wires above the fruiting wire used to guide growing shoots into the best location for receiving the maximum sun. Also called positioning or training wires.

H

HARVEST When the ripe grapes are picked. It is 70 to 140 days after berry set.

HARVESTING The process of picking the ripe grapes.

HYDROMETER A gauge used to measure specific gravity of grape juice to determine the percentage of sugar, or Brix.

I

ICEWINE Sweet dessert wine made from grapes allowed to remain on the vine well into winter, so that they freeze.

INTERNODE The portion of a shoot or cane between two adjacent nodes.

L

LATE-HARVEST WINE Sweet dessert wine made from grapes allowed to remain on the vine after ripening and sometimes infected with botrytis.

LATERAL SHOOT A shoot that develops at a leaf node off the main shoot. These laterals may reach several feet in length and need to be removed.

LEAF FALL The period when vines begin to go dormant in the late fall and early winter. Leaves turn red, yellow, and then brown, and drop off the plant.

LEAF PULLING The technique of taking leaves off the vine during the summer in order to expose the inner plants and fruit to the sun.

LOAM The ideal soil, rich in nutrients, with a balanced structure and plenty of organic matter.

M

MACROCLIMATE The climate in a broad geographic region.

MESOCLIMATE The differences in climate that occur within a macroclimate.

MICROCLIMATE The immediate climate that occurs in the vineyard.

MUST The liquid solution of crushed grapes.

N

NEMATODES Any of a wide variety of microscopic roundworms in soil. Some are beneficial and some are pests that feed on vine roots.

NODE The bulges in the shoot or cane from which buds appear. These buds develop into leaves, tendrils, clusters, or other buds.

P

PETIOLE The stalk that attaches a leaf to a stem.

PH A measure of acidity or alkalinity, measured on a scale from 1 (very acidic) to 14 (very alkaline). pH is measured both in soil and in grape juice.

PHOTOSYNTHESIS The process by which green plant tissues convert energy from the sun into food for the plant in the form of sugar.

PHYLLOXERA A root louse that attacks grape roots, causing stunted growth and vine death.

POST-VERAISON The period after veraison, when the grapes begin to soften and develop their final color.

POWDERY MILDEW The most serious and widespread disease in vineyards in terms of expense and loss of crop yield.

PRUNING The process of removing parts of the vine to prepare it for next growing season's pattern of growth.

R

REFRACTOMETER An instrument that measures the percentage of sugar, or Brix, in grape juice or must.

RENEWAL SPUR A spur that forms on a cane from the previous year's spur.

ROOTSTOCK The bottom part of a grapevine to which the upper part, or scion, is grafted. Some grapevines are grown on their own roots.

ROOT ZONE The area below ground where the roots grow.

ROW ORIENTATION The compass direction to which the rows in a vineyard are oriented.

S

SCION The upper part of the grapevine that is grafted onto a rootstock.

SHOOTS The green growth of the trunk, cordon or spur. Shoots will grow long if not trimmed back or pulled.

SHOOT DEVELOPMENT A shoot begins to develop at bud burst and begins to elongate as leaves unfurl at the apex. As the fruit ripens the shoot tip will slow down or stop growing.

SHOOT POSITIONING The process of training the direction of shoots up between the double wires of the trellis system.

SHOOT POSITIONING WIRE The double strand of wires above the fruiting wire into which shoots are trained.

SHOOT PULLING The process of taking off laterals and suckers from the grapevine.

SHOOT SPACING The distance between shoots on the cordons or arms of a more mature vine.

SHOOT THINNING See Shoot pulling.

SHOOT TRAINING Process of tying shoots to stakes and wires, and pushing them to grow between the shoot positioning wires.

SHOOT VIGOR The speed and strength of a grapevine's growth.

SPUR A short, fruiting portion of first-year growth (cane), which is pruned down to one to three nodes.

STOMATA Tiny pores under the leaf of the plant which allow plant respiration. Stomata are light-sensitive, so they close at night and open during the daylight hours.

SUCKER A shoot arising at or above ground level. A water sucker, or water shoot, is a shoot that will not bear fruit.

SUCKERING (ALSO KNOWN AS DE-SUCKERING) The process of removing shoots at or just above ground level and those arising from the trunk or spurs, usually water shoots.

TA See Acid.

TENDRIL A tentacle-like climbing structure that emerges from a node and attaches or wraps around adjacent shoots or wires.

TENSIOMETER A type of soil moisture sensor.

TIGHTENER (ALSO KNOWN AS TENSIONER) Devices used at the end post (sometimes both) of the trellis system to help pull the wires taut and keep them there.

TRAINING The process of training the growing vine on a trellis system to ensure a good mature vine structure.

TRANSLOCATION The process by which chemical materials and nutrients move through the vine. For example, sugar can be exported to either the shoot tip, grape cluster, root system, or other permanent parts, such as the trunk, for storage.

TRANSPIRATION The process whereby water, during sunlight hours, moves through the plant's system, up from the roots, and out through the underside of the leaves through the stomata.

TRUNK The center portion of the vine above ground.

VARIETY Any of a group of genetically identical plants within a species that have distinct, uniform, and stable characteristics, such as Cabernet Franc, Pinot Noir, Merlot, and others.

VERAISON The stage of a berry midway through its development, when it changes from hard to soft, and the color changes from green to its harvesting color (depending on the type of grape it will turn to black, blue, golden, green, pink, or yellow).

VERTICAL SHOOT POSITIONING (VSP) A trellis system that allows for the training of vines up from a cordon between sets of double wires.

WATER BUDGET METHOD A means of gauging when to irrigate based on how much water is present in the ground around the root system.

WATER STRESS Occurs when soil water supply is low and the weather is hot, sunny, windy, and of low humidity. The signs of water stress are when shoot growth stops and the plant looks grayish in color.

Helpful Conversions

INCHES	CENTIMETERS
1/8	0.3
1/5	0.5
1/4	0.6
1/3	0.8
2/5	1.0
1/2	1.25
1	2.5
2	5.0
3	7.5
4	10
5	12.5
6	15
7	18
8	20
9	23
10	25

FEET	METERS
1/4	0.08
1/3	0.1
1/2	0.15
1	0.3
2	0.6
3	0.9
4	1.2
5	1.5
6	1.8
7	2.1
8	2.4
9	2.7
10	3.0
20	6.0
30	9.0
40	12
50	15
100	30
200	60

POUNDS	TONS	KG
100		45
200		91
300		140
400		180
500	1/4	230
1000	1/2	460
2000	1	910

ACRES	HECTARES
1	0.4

Resources

Further Reading

Fox, Jeff. 1999. *From Vines to Wines. The Complete Guide to Growing Grapes and Making Your Own Wine*. North Adams, MA: Storey Publishing.

Jackson, David, and Danny Schuster. 2001. *Production of Grapes & Wine in Cool Climates*. Lincoln University, PA: Lincoln University Press.

Kourik, Robert. 2009. *Drip Irrigation for Every Landscape and All Climates*. Occidental, CA: Metamorphic Press.

Magarey, Peter A. 2000. *Field Guide to Diseases, Pests and Disorders of Grapes*. South Australian Research and Development Institute. South Australia, AU: Winetitles.

Ohlendorf, Barbara. 2008. *Grape: UC IPM Pest Management Guidelines*. Publication 3448. Richmond, CA: University of California Agriculture and Natural Resources.

Ortho. 2006. *All About Sprinklers and Drip Systems*. Hoboken, NJ: Wiley.

Smart, Richard, and Mike Robinson. 1991. *Sunlight into Wine: A Handbook for Wine Grape Canopy Arrangement*. South Australia, Australia: Winetitles.

University of California. 1992. *Grape Pest Management*. 2nd edition. Richmond, CA: UC IPM Publications.

Warrick, Sheridan. 2010. *The Way to Make Wine: How to Craft Superb Table Wines at Home*. Berkeley, CA: University of California Press.

Winkler, A. J., J. A. Cook, W. M. Kliewer, and L. A. Lider. 1974. *General Viticulture*. Berkeley, CA: University of California Press.

Online Resources

Canada Organic Regime
inspection.gc.ca/english/fssa/orgbio/stainte.shtml

Cornell University College of Agriculture and Life Sciences Viticulture and Enology Program Outreach
grapesandwine.cals.cornell.edu/cals/grapesandwine/outreach/index.cfm

Grapestompers
grapestomers.com

The National Organic Program
ams.usda.gov/nop/

National Sustainable Agriculture Information Service (ATTRA)
attra.ncat.org

Texas Winegrape Network
winegrapes.tamu.edu

University of California, Davis, Foundation Plant Services
fpms.ucdavis.edu

University of California UC Integrated Viticulture
iv.ucdavis.edu

University of Minnesota, Cold Hardy Grapes
grapes.umn.edu

University of Vermont, Cold Climate Grape Production
uvm.edu/pss/grape

Winemaker Magazine
winemakermag.org

Index

acid / acid balance / acidity, 160, 161–162, 170
Agrobacterium tumefaciens, 136
air root, 170
America grape, 46
ammonium nitrate, 116
Anagrus, 150
anthracnose, 84, 134
apex, 170
aphids, 120, 155
arm. *See* cordos
assassin bugs, 155
augers, soil, 72, 75
Aurora grapes, 41
average daily temperature, 30, 40, 85
azimuth, 170

Bacillus thuringiensis and *Bacillus popillae,* 149, 151
backpack sprayer, 132–133
Baco Noir grape, 46
bacterial diseases, 131, 136
Barbera grape, 46
bark, 97, 170
barley, 121
barrel aged wine, 43
barrel press, 168
bat guano, 115, 116
Beaujolais variety, 49
bees, for pollination, 120, 145
beetles, 147, 149, 155, 156
bench-grafting, 52, 170
Bermuda grass, 66
berry development / set, 29, 170
bindweed, 66
birds, 144, 139. *See also* netting
  control, 140, 141–142
  insectivorous, 149
black aphids, 120
blackberry, 66, 114
black measles, 134
black rot, 41, 46, 135
Black Spanish grape, 46
blando bromegrass, 121
blending and blending wines, 48, 49, 50, 170
bluegrass, annual, 121
Bobcat augur, 72
Bordeaux grapes, 44, 48, 49, 96
Bordeaux mixture spray, 130, 137, 139
borers, 84, 147
boron, 33, 115
botrytis / *Botrytis cinerea,* 135, 170
branch and twig borer, 147
British Columbia grapes, 41, 43
Brix, and measuring, 87, 88, 161–162, 163, 170
bromegrass, 121, 122
bud burst / bud break, timing of, 27, 29, 41, 44, 48, 49, 55, 85, 96, 170, 172
buds, and pruning / maintenance, 94, 98, 100, 103, 105
"bull" canes, 92
bunch rot, 41, 44, 48, 87, 135, 166
burclover, 120
burgundy grapes, 41

calcium, 33, 170
Cabernet Franc grape, 46, 96, 104, 174
Cabernet Sauvignon, 39, 40, 48, 96, 104, 138, 162
California, as prime location, 22, 28, 50, 51
California Department of Food and Agriculture, 40
Canada Organic Regime, 20
Canadian growing regions, 22–23
canes / fruiting canes, 92, 98, 170
canopies and canopy management, 19, 28, 92, 99, 100, 104–105, 171
  for disease and pest control, 126, 130, 134, 135, 138
Carignan, 138
caterpillars, 85, 155, 156
Cayuga grape, 41
Chancellor grape, 48
Chardonnay, 38, 40, 41, 42, 138

Chambourcin grape, 48
Chardonnay grape, 41, 139
Chelois grape, 48
Chenin Blanc grape, 28, 41, 138
chickens, 145, 154
chipper, for old canes, 93
chlorophyll, 33, 171. *See also* photosynthesis
chlorosis, 33, 171
clay soils, 30, 59, 110, 113, 120–121, 170, 171
climate, and differences in seasons, 15, 26–28, 40
clones, 38–39, 40, 41, 50, 171
clover, 119, 120, 122
cluster rot, 44
colder / cooler regions
  grapes for, 38, 41, 43, 44, 50, 43
  irrigation in, 111
Colombard grape, 40
compost / manure, 34, 68, 88, 99, 171
  for healthy soil, 114, 115, 116, 150, 153
  for weed control, 117
Concord grape, 48
construction tools, equipment, and supplies, 72–73
conversions, 176
copper and copper sulfate, as fungicide, 33, 130, 135, 137, 138
cordons, training and pruning, 92, 94–95, 96, 97, 98, 171, 172
Cornell University
  grape breeding, 39, 41
  pest management, 127
cover crops, 35, 64, 68, 84, 85, 86, 87, 88, 171
  for beneficial insects, 145
  for insect pest control, 149
  legumes as, 116
  for maintenance and water retention,108, 112
  for weed control, 117, 118–122
coyotes, 141
crimson clover, 120
Crown gall, 136
Cynthiana grape, 50

D

damsel bugs, 155
De Chaunac grape, 48
deer control, and fence, 64, 68, 86, 108, 140
Delaware grapes, 48
diatomaceous earth, 147, 149, 153
disease(s), 134–139. *See also* fungal diseases
  controlling / management, 20, 28, 85, 88, 93, 99, 126–127, 130–133
  resistance and susceptibility to, 41, 48, 49, 50, 51
  dormancy, 19, 28, 54, 82, 84, 92–93, 171
  maintenance and, 109, 112, 113
Dornfelder grapes, 48, 160
downy mildew, 48, 49, 136–137
drainage, 32, 50, 64, 119
drip irrigation and drippers, 57, 59–60, 64, 75, 79, 85, 109, 111. *See also* irrigation
  fertilizing using, 113–114
  post-planting, 82
drivers (tools), 72
drought-resistance and drought years, 109, 112
ducks, 145

E

early spring activities, 85
early summer activities, 86
earthworms, 118
Edelweiss grape, 41
electric fencing, 140
Ehrenfelser grape, 41
Elsinoe ampelina, 134
Environment Canada office, 26
equipment, construction, 72
erosion control, 26, 118, 120–122
esca, 134
eutypa / eutypa dieback / *Eutypa lata,* 86, 137
evapotranspiration, 59, 60, 171

F

fall activities, 88
fava beans, 68, 119, 120
feeding your vines, 116

fencing and fencing tools, 68, 72, 140, 141
fertilizing / fertilizer, 85, 89, 105, 113–114, 171
  organic, 115, 119
  for insect pest control, 151
fescue, 119
field grafting, 52, 171
Finger Lakes, New York, vineyard, 160
fleabane, 66
flea beetles, 147
flowers / flowering, 29, 55, 86, 104, 172
Foch grape, 48, 49
foliage. *See* leaves
foxes, 87, 88, 141
French drain, 82, 172
French wines, 30
Frontenac grape, 42, 43, 47, 49
frost-sensitivity, frost dates, and minimizing damage, 26–27, 40, 41, 85, 118
fruiting zone, 99, 142
fruit flies, 139
fruit set, 86, 99, 170, 172
fungal diseases / fungus, 28, 85, 86, 92, 108, 117, 126, 172
  spores, 130–131
  sprays for, 132
  treatments for specific, 134–138
fungicides, 131, 134–137

G

Gamay grapes, 47, 49
garden tortrix, 151
geese, 145
German grapes 41, 48. *See also* Gewürztraminer
Gewürztraminer, 28, 40, 42, 43
glassy-winged sharpshooter, 138, 148
goats, for weed control, 66
gophers and gopher control, 85, 86, 87, 117, 141, 142
grafting, 52
grape berry moth, 148
grape cane borer, 147
grape cane gallmaker, 149
Grape Pest Management (UC Davis), 127
grape pomace, 88
grape varieties, popular, 41, 43, 44, 46, 48–51
grapevines, buying, 39–40
grasses, 121–122
grasshoppers, 87, 149
Grenache, 40, 49
Grippler tool / Gripples, 72–73, 77–78
ground cover, 27, 86, 87, 108, 118, 119. *See also* cover crops for beneficial insects
growing regions in U.S., 22–23
growth and growth stages, 29
grow tubes, 64, 65, 74, 81, 82, 117, 141, 172
Guidnardia bidwellii, 135

H

hardiness, 52
harvesting and harvesting equipment, 88, 89, 98, 112, 164–165, 172
hedgerow, need for, 20, 145
herbicides, 117, 133
Himalayan blackberry, 66
hornets, 145
hotter climates, 24, 28, 43, 44, 50, 51, 99. *See also* average daily temperature
  irrigating in, 111
humid conditions, 38
hydrometer, 161, 163, 172

I

icewine, 28, 43, 44, 46, 89, 160, 172
insect(s)
  beneficial, 26, 118, 119, 145, 151, 155–157
  pests / vectors, 108, 138, 145, 147–154
insecticides, organic, 145–146, 152
installation timeline, 64–65
Integrated Pest Management (IPM), 20
iron, as nutrient 33
irrigation system, supplies and tools, 14, 56, 59–60, 64, 74, 79, 83. *See also* drip irrigation; water management
  for insect pest control, 151
  schedule, 109–111
Italian grape varieties, 46, 51

J

Japanese beetles, 86, 87, 149
Jefferson, Thomas, as vintner, 22

K

kaolin, 148, 153

L

lacewings, 150, 155, 157
lady beetles, 150, 155
lana vetch, 120
land, clearing, 66
late fall activities, 88
laterals / lateral shoots, and removing, 99, 100, 105, 171, 172
late spring activities, 86
late summer activities, 87
late winter activities, 84, 97, 105
layout, vineyard, 69
leaf fall, 29, 109, 111, 172
leafhoppers, 85, 87, 148, 150, 155
leaves
 and leaf pulling, 28, 99, 100, 101, 105, 173
 thinning, 87
 turning color, 88
legumes, 68, 116, 119
Lenoir grape, 46
Leon millot grape, 49
lignin formation, 33
lime-sulfur application, 84, 134
"little leaf disease," 33
loam soils, 30, 55, 59, 110, 120–121, 173
location considerations, 22–24
London rocket, 66

M

macroclimate, 22–24, 173
macronutrients (major nutrients), 33, 34
Madeleine Sylvaner, 28
magnesium, 33, 115
maintenance, vineyard, 19, 58, 126. *See also* irrigation
Malbec grape, 49, 96, 104
Malvasia, 43
management practices, 15, 27, 126–127
mantis, 155
manure. *See* compost
Marechal Foch grape, 49
maritime climate, 25
Marsanne grape, 43, 44
mealybugs, 84, 85, 88, 131, 153, 155
medics, 120
Merlot wine and grape, 39, 40, 47, 49, 96, 104, 174
mesoclimate, 23, 40, 173
metal post driver, 72
microclimate, 173
micronutrients, 33, 34
midspring activities, 85
mid summer activities, 87
midwinter activities, 84
mildew spores, 130
minute pirate bugs, 150, 152, 156
mites, 155, 156, 156. *See also* spider mites
molybdenum, 33
Mondavi, Robert, 30
moths, 148, 151
Mourvedre grape, 50
mowing, 117
mulches, 112, 117, 118 *See also* compost
Muscadine grapes, 38, 50
muscat grapes, 43
must, 173
mustard, 119

N

Napa Valley wines, 30
National Organic Program (NOP), 20, 131
National Sustainable Agriculture Information Service (ATTRA), 20
Nebbiolo grape, 50
neem oil, 149, 152
nematodes, 34, 52, 55, 131, 147, 149, 150, 153, 156, 173
netting protection from birds, 27, 87, 88, 142–143, 144, 149
New York, Long Island, winemaking, 25
New York Muscat, 28, 50

Niagara grape, 42, 43
Niagara region, Ontario, Canada, 23
nitrogen, 113, 114, 116, 118, 119, 121
  and nitrogen-fixing bacteria, 33
North American growing regions, 22–23
Norton grapes, 22, 50
N-P-K nutrients, 115
nutrient(s)
  deficiencies, 108, 116
  major soil, 33

O

Ontario, Canada, wines, 23
open canopy, 19
opossums, 141
Orange Muscat grape, 43
orange tortrix, 151
Organic Fertilizer Association of California, 115
organic fertilizers, 115, 119
organic matter in soil, 30, 34
Ortega grape, 43
owls and owl box, 141, 142, 144

P

Pacific Northwest climate, 28
parasites. *See* insect(s)
pathogens, soil, 34
peas, winter, 121
pesticides
  chemical, 21
  label, 133
  spray, 132–133
  use and disposal, 128–129
pests, vineyard, 20, 68, 85, 87, 88. *See also* insect(s)
  animal, 140–143
  control, 126–127
  maintenance and, 102, 108, 115, 117
petiole and petiole test, 113, 116, 173
Petit Sirah, 50, 138, 164
Petit Verdot, 96, 104
pheromone traps, 85, 146, 148, 149, 151
pH levels, 31, 33, 34, 35, 52, 69, 88, 160, 170, 173
pH meter, using 161–162
phomopsis / phomopsis cane and leaf spot / *Phomopsis viticola,* 86, 137
phosphorus, 33, 116
photosynthesis, 17–19, 29, 92, 116, 160, 173
phylloxera, 34, 52, 54, 55, 151, 173
picking knives and bins, 164–165, 167–168
Pierce's disease, 41, 46, 148, 138
Pinot Blanc wine and grape, 28, 43, 139
Pinot Gris (Pinot Grigio) grape, 44, 45, 139
Pinot Noir wine and grape, 23, 28, 38, 40, 47, 50, 139, 174
piperonyl butoxide, 146
planting, timeline, and depth, 64, 81
Plasmopara viticola, 136–137
pollination, 29
polypropylene mesh fence, 68
post-planting, 82
posts / post-hole digger / mid-posts / end posts, 64, 72–73, 75, 80
potassium and potassium deficiency, 33, 116
powdery mildew, 44, 48, 136, 138–139, 173
praying mantis, 149, 155
pre-planting, 67
prevention, pest and disease, 126–127, 138
propane flamer, 126
pruning / pruners, 28, 84, 92–93, 139, 164–165, 166, 173
  fifth year and beyond, 104
  first year, 94–95
  fourth year, 102–103
  second year, 97
  third year, 98
pyrethrins / pyrethrum, 147, 149, 150, 153
PVC pipes / risers, 59, 79, 171

R

rabbits, protection from, 74, 88, 141
raccoons, 87, 88, 140, 141, 142
rainfall toward end of growing season, 40
rapters. *See* owls
rats, 141
Ravat 51 grape, 44

red wines, 46, 48–51
reflecting tape, 142
refractometer, 87, 88, 161, 163, 173
renewal spur, 98, 100, 103, 173
Rhone grape varieties, 43, 44, 46, 50
Riparia gloire, 55
Riesling wine and grape, 23, 27, 40, 44, 45, 139
rootstocks, 30, 31, 38, 39, 52–55, 69, 173
  resistant to pests, 150, 151, 138
  vigorous, 105
root systems and growth, 30, 33, 85, 89, 109, 111, 116
rose clover, 120
rototiller, 66
Rougeon grape, 50
roundworms, 150
Roussanne grape, 44
rows, spacing and orientation, 14, 15, 25, 26, 28, 69, 70–71, 98, 173
Ruby Cabernet, 40, 51, 138
ryegrass, 119, 120

S

sample plan for small vineyard, 71
sandy soils, 25, 30, 33, 55, 59, 110, 113, 120–121
Sangiovese grape, 51
Sauvignon Blanc wine and grape, 28, 44
Semillion grape, 44
Schuyler grape, 51
scion, 38–39, 52, 173
seasons. *See* climate
seed dispenser, 119
Seyval grape, 44
shade obstacles, 26
sharpshooter pests, 85, 87, 138, 148
sheep, for weed control and manure, 67, 115
shepherd's purse, 66
shoots and shoot growth, cutting / training, 81, 86–88, 92, 94, 96, 97, 101, 104–105, 173–174
  irrigation and, 109, 111
silt, 30
slopes, considerations and preferences, 23, 26–27, 31, 69, 105
soil assessment, nutrition, and testing, 30, 32–35, 64, 113, 117
soil erosion, 118, 120. *See also* erosion control
soil pathogens, 34, 108
soil preparation, 68
soil qualities and texture, 31–32, 121
solarization, 64, 67
sour rot, 139
sowthistle, 66
spacing, between vines and rows, 15, 24, 69, 70–71
Spanish grapes, 51
spider(s), 145, 150, 151, 156
spider mites, 84, 85, 87, 152, 155
spinosad, 151
squirrel repellants, 141
sprays / spraying for. *See also* backpack sprayer
  animal control, 140
  disease control, 130–131, 132–133, 136
  insect control, 145
  weed control, 117
spring. *See* early spring; midspring; late spring
spurge, 66
spurs, 92, 94, 95, 96, 97, 98, 103, 104, 105, 174
stakes, grape, 80, 94
starlings, 141
stomata, 19, 174
storage and winemaking prep, 166–168
string trimmer, 117
Stylet-Oil, 152, 136, 138
subclover, 120
subterranean (French) drains, 82
sugar-acid balance, 160–162
sulfur and sulfur spray, 33, 84, 130, 135, 138, 140
summer climate, and tips for, 24, 28. *See also* early summer, midsummer, late summer
sunlight needs, 17, 23, 24, 26–27, 28, 31, 92
supplies, construction, 73
sustainability, 20–21
Sylvaner, 138
Syrah wine and grape, 28, 40, 51

T

TA (titratable acidity), 161–162, 170
temperate zones throughout the world, 21
temperature requirements for winegrowing, 19, 29. *See also* colder / cooler regions; average daily temperature
Tempranillo, 28, 51
tensiometer, soil, 60, 86, 92, 174
Texas hill country, 24
Texas port wine, 46
thrips, 85, 152, 155, 156
Togninia species, 134
tools, construction, 72
training the vines. *See also* pruning
  fifth year and beyond, 104
  first year, 94
  fourth year, 100–101
  second year, 96–97
  third year, 98
transpiration, 17–19, 174
traps
  animal, 141
  insect, 146
trellis system / design / supplies, 14–15, 28, 56–58, 59, 73
  building, 75–81
  modified, 105
trimming top growth, 100
triticale, 122
trunk, and trunk training, 94, 97, 174
turkeys, wild, 141

U

University of California, Davis
  grape breeding research, 39
  pest management, 127
University of Minnesota hybrid grapes, 43, 49
urea, 116
U.S. Department of Agriculture
  National Organic Program (NOP), 128
  weed killer, 66

V

veraison, 29, 87, 112, 136, 139, 160, 162, 174
Verdelet grape, 44
Verdelho, 28
vetches, 119, 120–121, 122
Vidal grape, 27, 46
vigor, and excess vigor, 52, 105, 171, 174
Villard grape, 46
vine mealybugs, 152
vine spacing, 69, 70–71
vineyard yields, 15, 16
vinifera grapes, 22, 38, 40
Viognier grape, 28, 44, 46
viral diseases, 131
Virginia wineries, 22
Vitis labrusca, 41, 43, 48
Vitus vinifera, 38
voles, 141
VSP (vertical shoot positioning) trellis, 28, 56–58, 172, 174

W

wasps, 150, 151, 156, 139, 165
water management / watering, 60, 101, 105, 108, 126. *See also* irrigation; drip irrigation
  first year, 109
  fourth year and after, 111-112
  post-planting, 92
  schedule, 110
  second year, 109–111
  third year, 110
water needs, 19, 26, 59
weeds and weed control, 66–67, 82, 84, 85, 86, 87, 108, 114, 126
  cover crops for controlling, 117, 118–122
  for insect pest control, 147, 150, 151
weevils, 153, 156
white flies, 127
White Riesling, 138
white wines, 41, 43, 44, 46
windbreaks, 26
wine making, 15

winter activities, 89, 92
winter regions and climates, 23, 27, 38. *See also* average daily temperature; midwinter; late winter
wire, fruiting and positioning, 57–58, 76, 87, 105, 172
  first through fifth years, 94, 96, 97, 98, 99, 100
wire, installing /wrapping, 77–79
wire cutters, clips, and vices, 72, 73, 76–79
wire fencing, 68
wire holes, drilling, 76
worms, 127

Xylella fastidoisa, 138, 148

zinc, 33
Zinfandel grape and wine, 51, 135
'Zorro' annual fescue, 122

Photo and Illustration Credits

Photos

Sanjay Ach: page 14
William Cochran, California Department of Food and Agriculture / Bugwood.org: page 148 (below)
Cristom Vineyard: pages 18, 87 (left)
Gyorgy Csoka, Hungary Forest Research Institute / Bugwood.org: page 151 (above)
Heritage Film Project: page 22
Maggie Hoffman: pages 158–159, 160
Nathan Jones: page 143 (above)
Tracy Ellen Kamens, Ed.D., DWS, CWE; co-founder, Grand Cru Classes: page 132
Kandschwar: page 163
K.Vins / Karen Bilancetti: page 116
Malcolm Lubliner: page 188
Maccheek / Oliver Macdonald: page 138
Everett Mar: page 143 (below)
John Morgan: page 83
Russ Ottens, University of Georgia / Bugwood.org: page 148 (above)
Phil Pellitteri, University of Wisconsin: page 151 (below)
Powers family: page 9
Donna Powers: page 111
Tom Powers: page 15
Steve Ryan: page 74
Mark Smith: pages 29 (second from bottom), 85 (right), 86 (left)
University of Michigan; David L. Hansen: pages 42 (upper right), 47 (upper left)
USDA ARS Photo Unit, USDA Agricultural Research Service / Bugwood.org: page 152

iStockphoto.com / Derek Abbott: page 29 (bottom)
iStockphoto.com / Adrian Baker: page 149
iStockphoto.com / BasieB: page 135
iStockphoto.com / Hilary Brodey: page 126
iStockphoto.com / Joseph Calev: page 157 (left)
iStockphoto.com / Rachell Coe: pages 84 (left), 166
iStockphoto.com / Dajas Design: page 112
iStockphoto.com / Susan Daniels: page 119
iStockphoto.com / Parker Deen: page 66
iStockphoto.com / Matteo De Stefano, page169
iStockphoto.com / Dierks Photo: page 117
iStockphoto.com / Digital Paws Inc.: pages 23, 27
iStockphoto.com / Elena Elisseeva: page 57 (left)
iStockphoto.com / Jon Faulknor: page 127
iStockphoto.com / flyparade: page 47 (lower right)
iStockphoto.com / fotolinchen: page 122 (center)
iStockphoto.com / Nathaniel Frey: page 84 (right)
iStockphoto.com / Susanne Friedrich: page 142 (above)
iStockphoto.com / Dave Huss / Graphically Speaking: page 24
iStockphoto.com / Gregor Inkret: page 133
iStockphoto.com / Andreas Kermann: page 35
iStockphoto.com / Mars Lasar: pages 60, 86 (right)
iStockphoto.com / Cristian Lazzari: pages 25, 167 (left)
iStockphoto.com / Sarah Lee: page 154
iStockphoto.com / Scott Leigh: page 31
iStockphoto.com / Nancy Louie: page 88 (left)
iStockphoto.com / Mac99: page 87 (right)
iStockphoto.com / Alistair McCaskill: page 65
iStockphoto.com / Roger McClean: page 122 (right)
iStockphoto.com / Kevin Miller: pages 2, 11, page 47 (upper right, lower left), 89, 124–125
iStockphoto.com / MireXa: page 93
iStockphoto.com / Wendy Olsen: page 34
iStockphoto.com / Pgiam: page 61
iStockphoto.com / Norman Pogson: page 99
iStockphoto.com / Liese Rose: page 140
iStockphoto.com / Frank Rotthaus: page 45 (right)
iStockphoto.com / Cornelia Schaible: pages 29 (second and third from top), 85 (left)
iStockphoto.com / Richard Semik: page 42 (lower right)
iStockphoto.com / Paul Senyszyn: page 122 (left)
iStockphoto.com / SimplyCreativePhotography: page 67

iStockphoto.com / swalls: pages 42 (top left), 53, 88 (right)
iStockphoto.com / Tanuki Photography: page 101
iStockphoto.com / Bernard Tigno: pages 90–91
iStockphoto.com / Alan Tobey: pages 29 (top), 118
iStockphoto.com / Rachael Towne: page 57 (right)
iStockphoto.com / Dariusz Urbanczyk: page 153
iStockphoto.com / Michael Westhoff: page 146
iStockphoto.com / Darren Wise: page 21
iStockphoto.com / Yula Zubritsky: page 168

Shutterstock / Brandon Bourdages: pages 12–13, 102
Shutterstock / Elena Elisseeva: pages 36–37
Shutterstock / Gyuszkofoto: page 92
Shutterstock / janr34: page 42 (below)
Shutterstock / Kletr: page 81
Shutterstock / David L. Lewis / Wishing Well Productions: page 98
Shutterstock / Denis and Yulia Pogostins: page 130
Shutterstock / Walter Quirtmair: pages 106–107
Shutterstock / Ravl: page 157 (right)
Shutterstock / TFoxFoto: page 45 (upper left)
Shutterstock / Tim Scott: page 97
Shutterstock / Ian Wilson: pages 62–63
Shutterstock / James Zandecki: page 167 (right)

Illustrations

Kate Francis: pages 17, 26, 38, 29, 56, 58, 59, 77, 78, 79, 80, 81, 94, 95, 96, 97, 99, 100, 103, 104, 144 (right)
David Jacobson: pages 60, 71, 72, 73, 74, 75, 165

About the Author

Tom Powers operates Alhambra Valley Ranch, a 58-acre sustainable family farm in western Contra Costa County, California. The farm is located on the urban edge, between Martinez, Orinda, Pinole, and Richmond. It produces grapes for wine, olives for oil and curing, and summer vegetables, including the best-tasting Cherokee Purple heirloom tomatoes in the world. On two hillside pastures Powers raises goats, sheep, llamas, and emus. Many of the animals are retired or rescued farm animals, but the sheep and goats have become another business element of the ranch.

Alhambra Valley was once filled with orchards and wine grapes before Prohibition (1919 to 1933), with such well-known farmers as John Muir and John Swett. Powers started his farm in 1998 and since 1999 has planted hundreds of backyard vineyards in Contra Costa and surrounding counties under the name of Diablo Vineyard Planting and Management. In 2005 he revived the historic grape-growing and wine-making tradition by starting a small commercial winery on his farm. His varietal wines and blends are sold directly to friends, neighbors, stores, and restaurants.

He has also added more farm products and processing activities, including a greenhouse operation so that he can get his prize Cherokee Purple tomatoes to market in May instead of July.

These activities constitute the fourth or fifth career for this former attorney, county supervisor, realtor, and consultant. However, Powers is from a long line of farmers, including his grandfather who started one of the first farms in the Imperial Valley in 1910.

Tom Powers was awarded Contra Costa County's Sustainable Farming Individual of the Year award in 2009. In addition to the ranch, he runs the Alhambra Valley Olive Oil Company and Alhambra Valley Wine Company and chairs the Brentwood Agricultural Land Trust Board and the Contra Costa County Agricultural Advisory Board.